AF739877

Intellectual Property, Community Rights and Human Rights

This book considers the issue of biodiversity in developing countries in relation to intellectual property rights, community rights and human rights. Drawing together a number of case studies of developing countries rich in biological and genetic resources including India, South Africa and Brazil, the book examines the access to plant genetic resources (PGRs) and their utilizations in the contexts of scientific- and commercial-oriented activities pursued both in the source and user countries. Exploring how community rights are protected in national biodiversity-related regulations and some international legal instruments, Marcelin Tonye Mahop also discusses the relationship between community rights and human rights in the context of biodiversity. The book looks at the issue of bio-piracy, asking whether this phenomenon should only be seen as a North–South clash, whereby biodiversity-rich countries of the Southern Hemisphere blame developed countries and their actors as its principal perpetrators. While recognizing that developing countries' actors play a role in this bio-piracy phenomenon, the book goes on to suggest alternative measures for the legal protection of community rights at the national level with the possibility of national and international enforceability.

Essential reading for students and scholars of intellectual property rights, biodiversity regulations and human rights, this book will also be of great value to researchers and members of professional organizations working in these subject areas. National and regional negotiators in the international processes dealing with the issues covered in the book will find it a useful tool that can help them to understand various facets of these processes.

Marcelin Tonye Mahop is a Researcher at the Centre for Commercial Law Studies, Queen Mary, University of London, UK.

Routledge Research in Intellectual Property

Available:

Re-thinking Intellectual Property
The Political Economy of Copyright Protection in the Digital Era
YiJun Tian

The Development of Intellectual Property Regimes in the Arabian Gulf States
Infidels at the Gates
David Price

Forthcoming:

A Politics of Patent Law
Crafting the Participatory Patent Bargain
Kali Murray

Intellectual Property in Global Governance
The Crisis of Equity in the Knowledge Economy
Chidi Oguamanam

The Law and Economics of Intellectual Property in the Digital Age
The Limits of Analysis
Niva Elkin-Koren and Eli Salzberger

Intellectual Property, Community Rights and Human Rights

The biological and genetic resources of developing countries

Marcelin Tonye Mahop

Routledge
Taylor & Francis Group
LONDON AND NEW YORK

First published 2010
by Routledge
2 Park Square, Milton Park, Abingdon OX14 4RN

Simultaneously published in the USA and Canada
by Routledge
711 Third Avenue, New York, NY 10017, USA

Routledge is an imprint of the Taylor & Francis Group, an informa business

© 2010 Marcelin Tonye Mahop

Typeset in Baskerville by
Taylor & Francis Books

All rights reserved. No part of this book may be reprinted or reproduced or utilized in any form or by any electronic, mechanical, or other means, now known or hereafter invented, including photocopying and recording, or in any information storage or retrieval system, without permission in writing from the publishers.

British Library Cataloguing in Publication Data
A catalogue record for this book is available from the British Library

Library of Congress Cataloging in Publication Data
Mahop, Marcelin Tonye.
Intellectual property, community rights, and human rights : the biological and genetic resources of developing countries / Marcelin Tonye Mahop.
p. cm.
Includes bibliographical references and index.
ISBN-13: 978-0-415-47942-4 (hbk)
ISBN-10: 0-415-47942-8 (hbk)
ISBN-13: 978-0-203-85298-9 (ebk)
ISBN-10: 0-203-85298-2 (ebk)
1. Biopiracy. 2. Biotechnology – Law and legislation. 3. Traditional ecological knowledge – Law and legislation – Developing countries. I. Title.
K1519.B54M34 2010
343.' 0786606 – dc22
2009043116

ISBN13: 978-0-415-47942-4 (hbk)
ISBN13: 978-0-203-85298-9 (ebk)

Foreword

In the past four decades, phenomenal developments in biotechnology and information technology have combined to push to the fore the great potentials and benefits of PGRs. It is now widely accepted that plants contain an immense store of chemical resources that could be used for a variety of purposes. The astonishing aspect of this unfolding phenomenon is the emerging realization by researchers and scientists that much of the knowledge of the diverse uses of plant resources are possessed by various persons and stakeholders in traditional communities where most of the plant resources have been grown and used for millennia. Thus, PGRs and traditional knowledge must of necessity be construed and understood as community assets. These assets are essential for maintaining the livelihoods of rural communities, and for sustaining myriad cultural and societal functions. However, the increasing scientific and commercial interests engendered by the huge potentials of bio-cultural plant resources has opened up fissures at various levels on the fair and equitable use of plant bio-cultural resources.

The emerging debate underscores the important point that in addition to scientific and commercial interests in the exploitation of these elements of bio-cultural diversity, other institutions and stakeholders, in particular, local communities, have relevant interests in PGRs and traditional knowledge. For several generations now, community assets from developing countries have been exploited by scientifically and technologically capacitated actors, sometimes legally but also illegally and generally unethically. Such exploitation activities are a cause of concern to local and indigenous communities ranging from overutilization and depletion of the resources, 'bio-piracy', lack of respect of communities' prior rights over their assets and failure to ensure community participation in decision-making processes and to provide them with fair benefits and compensation. Sadly, in the power structure which imbues each stakeholder with voice and capacity in deciding who gets what 'rights' in the debate regarding the exploitation of resources, local communities are often given the short shrift.

While myriad national, regional, and international regulatory instruments and regimes on plant resources have focused on the intellectual property rights dimensions of plant resources, the underlying issue of fairness and justice to local communities has not gained equal prominence. Local communities lack voice because their contributions to PGRs are not respected and considered worthy of bearing legal rights. This stems largely from an elitist conception of what constitutes knowledge and which knowledge is worthy of respect and legal protection. This condescending attitude towards local communities diminishes the millennia of scientific contributions of local communities while ignoring the palpable human-rights dimensions of the regulation of access to and use of community assets for scientific and commercial purposes and the management of and control over access to these assets by local and indigenous communities.

This book fills a significant gap in contemporary scholarship on the human-rights aspects and justice-based formulation of the rights of local communities to plant resources. In pulling together an analysis of a cross-section of case studies of exploitation of community assets and trespass upon their human rights and of national, regional and international intellectual property-related biodiversity instruments and human-rights instruments drawn from the Americas, Asia Pacific and Africa, the author has crafted an elegant, readable, and penetrating analysis of the human-rights dimensions of the lingering neglect and denial of the human-rights issues raised by the scientific and commercial exploitation of community assets. Unless and until we understand the neglect of local communities and exploitation of traditional knowledge of the uses of plants as human-rights issues, the quest for sustainable development will be a mirage. Those lessons still need to be learned and this excellent book is a good place to start.

Ikechi Mgbeoji
Professor of Law
Osgoode Hall Law School, Toronto, Canada

Preface

PGRs and traditional knowledge, defined in this book as 'community assets', are very important elements of bio-cultural diversity, attracting scientific and commercial interests nationally and internationally. However, these elements of bio-cultural diversity are also essential for maintaining the livelihoods of rural communities in the developing countries rich in these assets. Apart from and in addition to scientific and commercial interests in the exploitation of community assets, other entities or institutions hold genuine interests in respect of their relationships with PGRs and traditional knowledge. Indeed, access to and potential exploitation of community assets are regulated through national, regional and international biodiversity regulatory schemes which are administered by specific institutions. Such regulatory schemes are generally developed to reflect well known legally and non-legally binding international and regional environmental laws, treaties and protocols such as the 1992 Rio Convention on Biological Diversity (CBD), the 2002 Bonn Guidelines on Access to Genetic Resources, the Andean Community Common Regime on Access and Benefit Sharing, and the 2001 International Treaty on Plant Genetic Resources for Food and Agriculture (IT-PGRFA) of the Food and Agriculture Organization (FAO) of the United Nations (UN). Other international treaties impacting on PGRs and traditional knowledge more oriented towards the commercial exploitation of these elements are the intellectual property treaties such as the 1994 Agreement on Trade Related Aspects of Intellectual Property Rights (TRIPS) of the World Trade Organization (WTO) and the conventions of the International Union for the Protection of New Plants Varieties (UPOV) especially the UPOV 1991 Convention. Additionally, there are indigenous peoples and professional bodies' declarations developed as soft instruments which attempt to streamline access to genetic resources and traditional knowledge and to balance the interests of all stakeholders involved in access processes such as local communities, researchers, industries and politically motivated decision-makers. The institutions administering all of these treaties and the peoples working for them all have interests in one or more aspects pertaining to the regulation of access to, use of and sustainable management of community assets. Above all, there is a human-rights dimension in the regulation of access to and use for scientific and commercial purposes and the management and control over access to community assets. This human-rights dimension is indeed grounded on the community way of life and cultural association with their assets and is recognized in the regional and international human-rights instruments.

It is widely known that from the colonial era till date, community assets from developing countries have been exploited by scientifically and technologically capacitated actors, sometimes legally but also illegally and generally unethically. Such exploitation activities are a cause of concern to local and indigenous communities ranging from overexploitation and depletion of their assets, 'bio-piracy', lack of respect of communities' prior rights over their assets and failure to ensure community participation in decision-making processes and to provide them with fair benefits and compensation. All these concerns are human-rights related, and understanding that and addressing them as 'human rights' elements can go a long way to mitigating community concerns and establishing trust between local and indigenous communities and other actors and entities including scientific and commercial users, institutions administering key regulatory instruments and decision-makers working for these institutions in national, regional and international settings.

It is against this background that the idea to write this piece emerged and the product emerged. It is my strong and profound belief that a practical consideration and construction of human-rights principles in the regulatory instruments of access to and utilization of community assets is the way forward to making biodiversity-related regulatory approaches aligned with the interests of all actors. This belief stems from two pillars. On the one hand, despite the fundamental differences and technical and legal complexities inherent to the application of patents and plant breeder's rights systems, it is important to enable local and indigenous communities to understand how these two IP tools operate, the rational for scientific and commercial operators' pursuit and acquisition of them and the potential benefits that may ever acrue to local communities. All these will gradually help them to accept and value the intellectual property rights such as patents and plant breeders' rights of others as legitimate rights. On the other hand, it is critical that all aspects of access processes and scientific and commercial utilization of community assets reflect the human rights of local and indigenous communities. The approach of this book is to bring about this balance through the development and suggestion of ways to implement intellectual property based and human-rights oriented biodiversity regulatory measures at the national level. The book has focused its analysis on four countries – Brazil, India, Peru and South Africa – used as case study countries. However, it is at the core of these analyses the understanding that the regulatory approaches discussed be replicable and adaptable to various national contexts according to every country's peculiar circumstances.

Acknowledgements

In the course of writing this book I received support from many senior colleagues and intellectual property and biodiversity policy and conservation professionals to whom I am heavily indebted. First of all, I want to express my profound appreciations and gratitude to Dr Graham Dutfield who is a professor of international governance at the University of Leeds for his very early courage-boosting words once my proposal for the publication of this book was accepted by Routledge. Despite his hectic schedule, Prof. Dutfield was kept informed on the progress on this book and maintained his support to my ambition to complete the manuscript. I want to stress that Prof. Dutfield, who supervised my Ph.D. while he was in post at the Centre for Commercial Law Studies, Queen Mary, University of London, always said to me that it would be good to have my Ph.D. materials transformed into a book, but that this would require expanding the scope of enquiries, thus meaning a great deal of further research. This product reflects exactly his suggestions, as it has added a significant human rights and PBRs dimension to the narrow scope of community rights in relation to patents and biodiversity regulations discussed in the Ph.D. dissertation. In addition, I received invaluable support from Prof. Ikechi Mgbeoji of Osgoode Hall Law School of York University, Toronto, for his technical and moral supports throughout the development of this book, all of which were crystallized by his acceptance to write the foreword. I will also be recognizant of the moral and technical supports received from some colleagues at the Centre for Commercial Law Studies (CCLS) at Queen Mary, University of London, who were informed of my plans to publish this book and of the acceptance of my proposal by Routledge and who were consulted as the writing-up process evolved. In particular, I want to express my sincere gratitude to Malcolm Langley, the Librarian of the Intellectual Property archive of the Centre for Commercial Law Studies, Queen Mary, University of London for allowing me to use the intellectual property archive.

Within the realm of biodiversity policy and conservation professionals, I here want to mention the support from colleagues of the African Conservation and Development Foundation, especially Dr Daniel Pouakouyou; colleagues from the African Centre for Biosafety with whom every now and again I am involved in research on bio-piracy in Africa; colleagues from the ABS (Access and Benefit Sharing) Capacity Development Initiative for Africa (www.abs-africa.info), in which I contribute as an African professional on biodiversity and intellectual property policy issues and especially on ABS issues.

These acknowledgements will be incomplete if I do not stress the moral support and joy provided to me by my family. All the way through, my wife Arlette Roseline Noutang Bassa Tonye, my two sons Ian Dimitri Mahop and Wilfred Mongo Tonye, and my mother Ngo Biya Anne, have provided me with those invaluable moments of relaxation, during the hectic periods of research and writing up.

Last but not least, I have very special thoughts to my late father, sister and godfather who made me the man and professional I am today but did not live to see these achievements.

Abbreviations

ABS	access and benefit sharing
ACB	African Centre for Biosafety
A(B)CHPR	African (Banjul) Charter on Human and Peoples' Rights
ACHPR	African Commission on Human and Peoples' Rights
AU	African Union
BIO	biotechnology industry organization
BIS	balanced information sharing
BMA	British Medical Association
CBD	Convention on Biological Diversity
CBD-COP	Convention on Biological Diversity Conference of the Parties
CBIPRs	Community Based Intellectual Property Rights
CBO	community based organization
CBPR	community based property right
CCM	cultural consensus model
CCP	code of civil procedure
Cefic	European Chemistry Industry Council
CICPR	community intellectual and cultural property rights
CIEL	Center for International and Environmental Law
CIG	common initiative group
CIMMYT	International Maize and Wheat Improvement Center
CIP	International Potato Center
CITES	Convention on International Trade in Endangered Species
CONAGEBIO	National Commission for the Management of Biodiversity (Costa Rica)
CPG	common policy guidelines
CSIR	Council for Scientific and Industrial Research
DEAT	Department of Environmental Affairs and Tourism (South Africa)
DSM	Dispute Settlement Mechanism
DTI	South African Department for Trade and Industry
EDV	essentially derived varieties
EIA	environmental impact assessment

EPC	European Patent Convention
EPO	European Patent Office
FAO	Food and Agriculture Organization (of the United Nations)
FAO-IU	International Undertaking of the Food and Agriculture Organization
FDA	Food and Drug Administration (United States)
GBS	global bio-collecting society
GREEN Foundation	Genetic Resource, Energy, Ecology and Nutrition Foundation (India)
HRC	Human Rights Committee of the ICCPR
ICC	International Chamber of Commerce
ICCPR	International Covenant on Civil and Political Rights
ICESCR	International Covenant on Economic, Social and Cultural Rights
IDCID	Institute of International Trade Law and Development (University of São Paolo, Brazil)
IGC	Intergovernmental Committee
IIFB	International Indigenous Forum on Biodiversity
IKS	indigenous knowledge systems (South Africa)
ILO	International Labour Organization
INDECOPI	Office of Inventions and New Technologies (Peru)
IP	intellectual property
IPR	intellectual property right
IR	international regime
IRABS	International Regime on Access and Benefit Sharing
ITPGR	International Treaty on Plant Genetic Resources
IT-PGRFA	International Treaty on Plant Genetic Resources for Food and Agriculture
IU	International Undertaking
IUCN	International Union for Conservation of Nature (or the World Conservation Union)
JPO	Japan Patent Office
MAS	marker assisted selection
MEC	Member of Executive Council (for environmental affairs at the provincial level in South Africa)
MoU	Memorandum of Understanding
NBA	National Biodiversity Authority
NCE	new chemical entities
NTFP	non-timber forests product
OAS	Organization of American States
OAU	Organization of African Unity
PBR	plant breeders' right
PGRs	plant genetic resources
PGRFA	Plant Genetic Resources for Food and Agriculture

PIC	prior informed consent
PVP	plant variety protection
QPM	Quality Protein Maize
RIG	resources interests group
SANBI	South African National Botanical Institute
SANBI	South African National Biodiversity Institute
ST	scheduled tribes (India)
STRC	Scientific, Technical and Research Commission
STRC-OAU	Scientific, Technical and Research Commission of the Organization of African Unity
TK	traditional knowledge
TRIPS	Agreement on Trade Related Aspects of Intellectual Property Rights
TRRs	traditional resources rights
UDHR	Universal Declaration on Human Rights
UN-DRIP	United Nations Declaration on the Rights of Indigenous Peoples
UNPFII	United Nations Permanent Forum on Indigenous Issues
UPOV	International Union for the Protection of New Varieties of Plants
USPTO	United States Patent and Trademark Office
WIPO IGC	World Intellectual Property Organization Intergovernmental Committee on Genetic Resources, Traditional Knowledge and Folklore
WIPO PCT	World Intellectual Property Organization Patent Cooperation Treaty
WIPO	World Intellectual Property Organization
WTO	World Trade Organization
WTO-DSM	World Trade Organization Dispute Settlement Mechanism

1 Setting the scene

Focus, main themes and broader objectives

As indigenous peoples, the protection (legal or otherwise, such as through voluntary instruments and professional bodies' declarations) of 'community rights'[1] at the national level in countries rich in biological and cultural diversity, needs to follow the path of a human-rights approach.[2] This suggestion is based on the recognition that community rights are crafted in many human rights instruments, international and regional, legally and non-legally binding (such as declarations), which should be taken into account in national policy-making. Unfortunately, in respect of national policy-making processes aimed at protecting community rights and addressing their concerns about access to and use of biological and genetic resources, and traditional knowledge (TK) in scientific and commercially motivated activities, there is often a lack of integration of human-rights considerations in the process. Furthermore, law and policy-making aimed at addressing the concerns of local communities are usually pursued in a very fragmented and piecemeal fashion at the national level, with no practical cooperation among the various actors and government institutions involved in the various aspects of management and regulation of community assets. As a consequence of this unpractical approach, there is no comprehensive and workable regulatory framework in force anywhere in the countries endowed with significant richness in bio-cultural diversity which can be viewed as providing protection to community rights from a human-rights perspective and addressing the various concerns of traditional and local communities – one of which being the issue of bio-piracy – over access to and use of their assets[3] by technologically advanced and capacitated users.

Ironically, to address their interests, technologically capacitated users of the assets of traditional communities use other weapons in their possession, such as the application of modern intellectual property rights (IPRs) such as patents or PBR systems over the outputs of their research endeavours based on the raw assets taken from local communities or their traditional lands. Analyses contained in the following sections and chapters are based on the realization and acknowledgement that IPRs and community rights are, in a way, all linked to each other within the framework of existing human-rights instruments. Indeed, even though human rights and IPRs are fundamentally dissimilar regimes because they pursue different aims, it is nevertheless the case that IPRs and community rights are all encapsulated in a number of international and regional human-rights treaties and recognized by international bodies such as the UN (more on this in Chapter 4). Despite this obvious linkage between human rights and IPRs, current regulatory avenues such as the ABS legislations (or biodiversity regulations) being established at the international and national levels and the voluntary mechanisms in the form of, for example, some declarations of professional bodies aiming to address the concerns and expectations of rural communities, fail to incorporate human-rights considerations from their development through to their implementation. Perhaps this situation is due to the lack of involvement of human-rights experts in these law- and policy-making processes. One of the consequences of this failure is that the weight and consideration given to IPRs in the end products from these law-making processes generally outweigh community rights and interests, perhaps due to the significant influence of corporate actors in these processes, who after all are not ready to lose, see themselves weakened, or have undermined all the benefits that they currently enjoy from strong consideration of IPRs. There seems to be profound and arguably obvious resentment on the part of traditional communities and actors sympathetic to their concerns that, indeed, in respect of policy development targeting issues of access to community assets and use of them in scientific and technology-intensive processes, IPRs of corporate and technologically capacitated actors have more influence than the community rights of local and indigenous peoples who are also traditional holders and custodians of biological resources and associated TK. For example, while corporate and research operators enjoy all the benefits arising from protecting their plants and TK-based research and development outputs by patents or PBRs including the associated financial benefits and the bargaining leverage, indigenous and local communities are being prevented from enjoying their own social and cultural benefits arising from their traditional rights over the very assets that formed the bedrock of technologically based research and development endeavours. Some of these rights that communities are arguably prevented from enjoying by PBRs are the rights to save and reuse planting materials for the next farming season and the rights to participate in decision-making processes pertaining to access to and the use of their assets at all stages of modern research and development processes.

How then should human rights come into play in an attempt to balance the interests of technology holders and modern users of the assets of traditional communities, the concerns of local and traditional communities over access and use of their assets, modern intellectual property (IP) over the application of the relevant outputs? This question is at the centre of the discussions and arguments running through the following sections and chapters of this book. Our point of departure is that in every instance involving access to community assets and use of them in profit-making technology-intensive processes with the application of patents and PBRs, we argue that stakeholders need to address the expectations of local communities from a human-rights perspective. This approach does not necessarily mean that the financial and other beneficial expectations of technology-holders would be affected negatively if they adopted a human perspective in the course of their actions. Rather, such an approach will more likely have no effect on such beneficial expectations of technology holders, but will instil and strengthen communities' trusts in the activities of technology holders, setting the stage for sustainable and more mutually beneficial relationships between local communities and the users of their assets. The human-rights considerations will be crafted in workable policy measures that will comprehensively integrate community rights and interests as well as the goals and objectives of the IPRs.

Using four carefully selected jurisdictions, namely Brazil, India, Peru and South Africa, this book focuses on suggesting some workable, comprehensive and enforceable policy measures embedded in regulations aimed at protecting the rights and addressing the concerns and expectations of traditional communities without undermining the goals and objectives of IPRs (patents and PBRs) in the context of access to community assets, use of them in scientific and commercially oriented activities. The book does so by decrypting how modern exploitation of community assets in research and development processes and the application of patents and PBRs, in line with the concept of bio-piracy, typically bring disrespect to community rights, which are recognized in regional and international human-rights law and processes. There is no intention here to blindly blame corporate and research institutions practices for bio-piracy. The subsequent bio-piracy-related accounts are objectively analytical in nature. They refute the bio-piracy concept when a specific case is not grounded on convincing evidence and, therefore, aim to oppose any non-evidenced or blatantly anecdotal negative impact of patents and PBRs on the rights, interests and expectations of indigenous and local communities. It must however be stressed that analyses carried out in this book are not strictly limited to the four selected countries. Where deemed necessary, the book will refer to cases and examples from other countries and areas rich in bio-cultural diversity, to the extent that such examples are relevant to the main themes discussed herein.

As it transpires from the above, the main themes discussed in the book are:

- Community rights. The rights of local and indigenous communities will be defined and discussed within the frames of the various laws and policies covered in this book.
- Human-rights instruments to the extent that they address community rights and concerns.
- Biodiversity-related regulations with special emphasis on the extent to which they address community rights, interests and expectations.
- IPRs, in particular, patents and plant variety protection (PVP) systems.
- Scientific and commercially oriented exploitation of community assets with the application of patents and PBRs in the protection of the relevant outputs – with an eye on the concept of bio-piracy.

The objectives of the book are among others:

- To examine cases of access to PGRs in Brazil, India, Peru and South Africa and their utilizations in scientific and commercially oriented activities pursued both in the source and user countries, noting that the source country may also be a user of the resources in question.

- To discuss the biodiversity-related regulations, national, regional and international instruments on access to and use of the biodiversity components. Rights-related issues such as prior informed consent (PIC), participation of local communities in the relevant decision-making processes and issues of enforcement of community rights both at the national and international level will be discussed.
- To discuss community rights as human rights. On the one hand, this discussion will look at how community rights are protected in international and regional human-rights-related instruments. On the other hand, the discussion will address the extent to which national policy processes embrace community rights as human rights as crafted in international instruments and recognized by international bodies.
- To discuss IPRs systems, especially patents and PBRs, their goals and objectives and the extent to which they address the interests and expectations of local communities and modern users of community assets. This discussion will be guided by specific cases as it intends to address the impact of some allegedly claimed bio-piracy.
- To suggest alternative and comprehensive policy and regulatory measures at the national level that are workable and convenient to address the interests and expectations of local communities from a human-rights perspective. Under this objective, the wider, international applicability of the proposed measures and their national and international enforceability are examined.

After setting the stage of our investigations, the following section attempts to unravel the interlinkages of the key themes discussed in the book, in a conceptual framework.

Conceptual framework: the nexus community rights, human rights and IPRs

The focus of this book is on exploring and suggesting practical avenues for the legal protection of the rights of rural communities at the national level in four selected countries, namely Brazil, India, Peru and South Africa, through comprehensive and workable IP-based biodiversity regulatory measures from a human-rights perspective. Our approach is to merge both IP systems (especially patent and PBR systems) and environmental and biodiversity-related regulations in a comprehensive framework, which encapsulates the concerns and expectations of rural communities as human-rights-related concerns. The concerns of rural communities about access to, acquisition of and use of their assets relate to, among other issues: the sustainability of their assets which occur in their culturally and traditionally owned lands; their PIC and rightful benefit-sharing expectations in respect of scientific and commercial use of their assets by scientific and commercial operators; and their rightful concern related to lack of participation in decision-making processes in which their assets are the core elements sought for eventual scientific and commercial exploitation. Generally speaking, various IP devices can be applied in the protection of technologically processed outputs based on the exploitation of the assets of traditional communities, despite the fact that these devices may as well be used to address the concerns of local and indigenous communities over access to corporate, scientific and commercially motivated use of their assets. For example, it has been well documented that copyrights, trademarks, geographical indicators and trade secrets can be used to address the concerns of rural communities about the commercial use of their intangible (TK, practices, dances and songs) and tangible assets (PGRs, artefacts).[4] However, this recognition is unfortunately not associated with practical cases as evidence of the way that these devices can be seen to be useful to traditional and local communities. Rather, there are increasing concerns that some IP devices are to the contrary harmful to local and traditional communities regarding the way they are applied in scientific and commercial exploitation of their assets. This book has chosen to focus on patents and PBRs as they represent the IP devices most used by scientific (increasingly academic researchers) and commercial users to secure private rights over the outputs deriving from their creative minds based on the exploitation of PGRs, TK and practices of local and indigenous communities.

The workable IP-based regulatory measures proposed and discussed subsequently address the concerns of rural communities and national interests, from a human-rights perspective, because, as we demonstrate below, community concerns are human-rights based. The proposed regulatory measures are framed to cover community rights over PGRs and TK, both the tangible and intangible assets. Indeed, a significant amount of anthropological studies stress the difficulties of separating these two components of the bio-cultural assets of traditional communities.[5] Moreover, an assessment of the relationships between local communities – who are the traditional custodians of PGRs and TK – and commercial and scientific users of these assets is additional evidence of the intricate linkages between the tangibles and the intangibles within the framework of exchange of these assets.[6] This book is one of the critical steps made towards the development of a comprehensive and integrated human-rights-based approach to protecting community rights and addressing their concerns. It is grounded on the practicality of the recognition of the linkages between PGRs and traditional germplasm (the tangibles) and the associated knowledge and practices (intangibles) of local and indigenous communities and traditional farmers in the developing countries.[7]

Another important approach adopted by this book relates to the manner in which it elaborates and considers the rights of rural communities that require legal protection. For many communities living in rural areas in biodiversity-rich countries across the world, PGRs and TK are generally owned in a communal manner. Using a model called the cultural consensus model (CCM) empirical studies provide strong evidence about how ethno-botanical knowledge is shared among communities from the same and neighbouring villages.[8] Within the framework of this model, the concept of TK is approached in its broadest sense. It encapsulates this type of knowledge which is held and nurtured by rural communities including but not limited to their traditional medicinal knowledge, traditional environmental-management knowledge, traditional farming and agricultural knowledge and the tangible resources to which such knowledge is associated. Taking all the components of community assets together as described within the realms of the CCM, this book's approach is that the associated rights of local and indigenous communities discussed herein should be broadened to community intellectual and cultural property rights (CICPR), because it provides a better understanding of the depth of the concerns and expectations of local communities.

The concept of CICPR over PGRs and associated TK and practices is therefore at the heart of our investigations. Although the major trend is to discuss community rights as communal rights, we do not totally ignore 'individual rights', although this dimension is not purposely argued for in the proposed regulatory measures. Indeed, as it emerges in a substantial amount of literature, in the Indian context for instance, some forms of knowledge at the community level are either owned individually or are only held by a particular and identifiable family lineage.[9] In the case of traditional medicinal or farming knowledge, it may be the case that just one individual or one family in a wider community masters the specific know how about the utilization of the medicinal plants or seeds (farming meterials). This individual or family will share this knowledge and the associated technical know how with community members under specific circumstances, for example when providing treatment to a patient within the community or if a member of the wider community typically requests guidance about how to plant the seads. Of course, this community approach in sharing resources, traditional innovation and practices weakens the concept of individual, private or family ownership rights at the community level, especially in African farming communities. According to one African IP expert, Andrew Mushita of Zimbabwe:

> In the African context, customary law is applied. It does not recognize private proprietary rights but rather community resources rights. All resources belong to everyone and they are regulated by the community's cultural and local knowledge systems and practices. In this sense, farmers have exchanged seeds among themselves since time immemorial, passing from neighbor to neighbor, mother to daughter, mother-in-law to daughter-in-law or even across villages and communities. Even labor is shared for such activities like land preparation, ploughing, planting, processing or threshing and harvesting crops. Land ownership has always been collective with individuals having user-friendly rights to land.[10]

From an anthropological perspective, Michael Brown discussed the difficulties of categorizing the traditional and cultural knowledge among indigenous communities as being either individually or communally owned. In investigating ownership issues in respect of the cultural designs and the Australian aboriginal ownership systems, the question addressed by Brown is whether, for example, such designs are owned by the artist, the clan of community, or some combination of the two.

Michael Brown also discusses the struggle of indigenous communities vis-à-vis the exploitation of their TK and their traditional lands within the framework of bio-prospecting activities. One specific matter Brown tries to address in his book in respect of ownership of knowledge as a traditional asset at the community level is whether such knowledge is owned by the individual who created it or such knowledge is just part of the culture of the whole community, thus needing to be considered as a communal asset. There is no clear-cut answer to this question, though Brown provides suggestions about how to deal with such issues in a pragmatic manner.[11] The approach to community rights pursued in this book as communal rather than individual or private rights over PGRs and TK is typically rooted on the concept of community resources that is based on the communal system of control and regulation over such resources at the community level, whether they are owned individually or collectively.[12]

The first step to securing effective participation of rural communities in the relevant decision-making processes pertaining to access and subsequent use of their assets is achieving a good understanding of the way in which communities are represented in such processes and how they provide their PIC. As we will reveal in the context of the specific South African case in Chapter 3 below, community representation and/or consultation in access processes is poor. However, from a practical point of view and in respect of the need for community consent within the framework of access mechanism at the national level, it is not the consent of every single member of a given community that will be required under the regulatory measures discussed in the final chapters. Our intent is to argue for community empowerment, which can only be achieved through their full and effective participation in the relevant decision-making processes through their traditional leaders or their elected members when they are organized in community based organisations (CBOs) or in resourses interests groups (RIGs) – which is a community group focusing on the sustainable use of some specific resources. Generally speaking, such community-based institutions are already proliferating in rural areas. Community rights of participation in policy development and decision-making processes on environmental issues affecting their life will be realized, if policy-makers work with existing local institutions. They will indeed be doing so in confidence, knowing that the leaders of these local institutions were elected by local communities, and thus represent these communities.

However within the framework of community participation, to be able to provide proper PIC, the communities will require some training and assistance from established civil society organizations such as Biowatch or the ACB in the case of South Africa. Expert policy analysts and environmental lawyers from these organizations have experience in interacting with and providing training to the locally established community institutions. Such training will aim at improving community awareness about the scientific and potential commercial value of their assets and the legal implications for giving their assets away during negotiations, one of those being the issue of enforcement of their rights through the formal judicial system. The involvement of expert lawyers to assist local communities during such negotiations coupled with their training will enable the communities to weigh up the implications in terms of potential gains or losses that will be associated with the acts of saying 'yes' of 'no' during such negotiations.

As it emerges from the foregoing, despite an obvious emphasis on patent rights due the accessibility to cases, farmers' rights[13] over their traditional germplasm against PBRs[14] are among the key issues discussed herein. The proposed regulatory measures are very relevant in situations of inappropriate access by plant breeders to farmers' traditional germplasm. In these situations, access to community assets raises identical concerns to rural farming communities including their exclusion from the decision-making processes through which plant breeders are authorized to access traditional germplasm, for example, via the extension services of the Ministries of Agriculture, the improvement of their germplasm through research and modern technological tools by plant breeders and possibility for using patents and PBRs to acquire legal private rights over the improved seeds. Furthermore, there may be commercialization of the outcomes of further research carried out on the germplasm without the consent of traditional farmers who provided the raw materials in the first place and without or with no proper compensation to local farmers. Not least, farmers' rights to reuse seeds during the next farming season may be affected by PBRs if farmers had used improved and legally protected seeds in the previous season. Let us briefly address farmers' rights *versus* PBRs in light of the provisions of the 2001 International Treaty of Plant Genetic Resources for Food and Agriculture of the United Nations Food and Agricultural Organization (FAO IT-PGRFA).[15] Having complied with the requirements for patentability, an applicant will be granted a patent over biodiversity-based inventions and will control the use of such inventions and might even make some revenue from it without giving any compensation to rural communities who provided the raw assets in the first place. Similarly, although the FAO IT-PGRFA provides for farmers' rights in its Article 9, it is not possible for farmers to stop plant breeders from claiming IPRs on the derivatives from such materials accessed through the multilateral system,[16] if they prove that their claims are based on the derivatives that are no longer in the form that the materials were received originally.[17] Rural communities' concerns about the fact that patent rights contribute to the theft of their medicinal plants and associated knowledge by academic and commercial bodies are identical to farmers' concerns and perceptions that PBRs contribute to 'theft' of their traditional germplasm by plant breeders.[18]

So far, I have tried to delineate the boundaries of the arguments pursued in this book in respect of community rights and IP acquisitions (patents and PBRs) over the derivatives of the resources (biological and genetic resources and associated TK) upon which traditional communities lay (traditionally rooted) rights-related claims. At this point the obvious question that needs attention is how human rights actually come into play in the whole equation regarding the legal protection of community rights, taking into account the role of IPRs in the exploitation process (scientific or commercial) of community assets over which communities lay their rights-related claims (addressed below). Indeed, there are a number of human-rights instruments (conventions and declarations) concluded under the aegis of the UN. While some of the instruments are not specifically aimed at indigenous and local communities' rights-related concerns over their assets, at least one of them is tailored to the peculiar circumstances and realities of indigenous and local communities. These human-rights conventions include, but are not limited to, the 1948 Universal Declaration on Human Rights (UDHR), its related and legally binding covenants, the 1966 International Covenant on Civil and Political Rights (ICCPR),[19] the 1966 International Covenant on Economic, Social and Cultural Rights (ICESCR),[20] and the 2007 United Nations Declaration on the Rights of Indigenous Peoples (UN-DRIP). As we will be demonstrating in the subsequent chapters, the major rights-related claims of traditional communities over their assets, including right of participation to decision-making processes, right of access and use of their assets, rights of PIC, (importantly) right to traditional lands and territories and right to self-determination, are all catered for by the above-mentioned legally and non-legally binding human-rights instruments. Alongside the UN-related treaties, other international organizations such as the International Labour Organization (ILO) have also established treaties pertaining to indigenous peoples' issues, which are relevant to our arguments and will thus be discussed subsequently.[21]

In its resolution 2000/7, the United Nations Sub-Commission on Human Rights acknowledges the actual and potential conflicts between the implementation of the TRIPS Agreement and the realization of the social and cultural rights contemplated in human-rights frameworks. In other words, this resolution recognizes the actual and potential conflicts between IPRs and human rights. Nonetheless, the same resolution in Article 1 endorses IPRs among the categories of rights covered by two international human-rights instruments, notably Article 27(2) of the UDHR and Article 15(1) of the ICESCR. This recognition literally brings IPRs and community rights within the realms of international human-rights law, despite the fact that human-rights and IP-rights regimes pursue fundamentally different aims. From our perspective, this recognition should work as a strong push for policy-making initiatives to make IP less detrimental to community rights. Rather, IPRs over derivatives from the assets of local communities should be grounded on and complementary to community rights if they are to achieve their goals and objectives in an ethical manner. In practical terms, there should be an adoption of a human-rights approach by actors in their pursuit of patents and PBRs. Such an approach needs to be crafted in enforceable regulatory measures. Such measures must require applicants for patents or PBRs in order, for instance, to take into account the concerns of rural communities before filing their applications for these private rights. In so doing, it is likely that community complaints over bio-piracy will be minimized and the sustainability and availability of their assets will be ensured in the interest of humanity. The proposed measures will be discussed and elaborated in the subsequent chapters. However, before we get to these substantive regulatory measures, let us present the subject matters at the core of our analyses. Thus, the following section discusses the peculiar circumstances of traditional societies alongside their assets. This section further explains how these societies conserve and use their assets in a sustainable manner.

Biological and cultural diversity conservation and use

Traditional societies, whose survival depends the most directly and heavily on biological resources, appear to be concentrated in areas of biodiversity richness. Tropical areas host 50–90 per cent of the world species and 1,400–2,500 ethno-linguistic groups, among the estimated 4,635 ethno-linguistic groups of the 225 eco-regions of the global 200 (a concept developed by the World Wildlife Fund (WWF) that attempts to identify a set of eco-regions whose conservation would achieve the goal of saving a broad diversity of the earth's ecosystems).[22] That these 4,635 ethno-linguistic groups represent 67 per cent of the total number of ethno-linguistic groups worldwide (6,867), demonstrates the link between cultural and biological diversity, in turn suggesting that a threat to one is likely to impact on the other.[23] Unfortunately, as we will see below, countries or areas rich in biological and cultural diversity generally lack the most favourable means and capacities optimally to exploit and conserve their resources.

In order to sustain the components of biodiversity and their usefulness to humankind, a number of strategies have been formulated in tropical areas. Alongside conservation projects funded by Western-based international organizations, more traditionally rooted strategies based on local communities' understanding of the functions of ecosystem are also attempted.[24] Considering the role that local communities play in the context of conservation and the sustainability of biodiversity, some commentators have argued that biodiversity conservation and the exploitation of its components would be most successful if they address the concerns of such communities and ensure their participation in the relevant decision-making processes.[25] Let us assess this in respect of the lifestyle of the Pygmies who live in the tropical forests of Central Africa. As the most ancient hunter-gather societies of the rainforest of the Congo Basin, the lifestyle of the Pygmies can be used as a model in an attempt to explore the issues of sustainability in relation to the livelihoods of rural societies.[26] There is no intention here to suggest that the natural resource-management practices of the Pygmies, based on their wealth of traditional and ecological knowledge, are unquestionably unsustainable. However, it may well be the case that, similar to the approaches employed elsewhere, such as in Tanzania,[27] a sound inclusion of the traditional management practices of the Pygmies of the Congo Basin in the activities of modern conservation initiatives working in this area may well yield successful results in terms of biodiversity conservation and sustainable development. Such trust in the Pygmies' traditional management practices may then be based on the recognition that, despite their permanent access to and heavy reliance on forest resources, the latter do not get depleted, this indicating some degree of sustainability of the forest management practices of the Pygmies.

In a wider dimension, it is reported that throughout Africa, conservation and traditional management practices of rural communities are regulated by local and culturally rooted rules, otherwise known as customary laws, which include fines and sanctions for wrongdoers. For example, in the Miombo region in southern Africa, the cutting of fruit trees growing around a specific water spring is prohibited. Sacred groves and other forest stands are maintained for important ceremonial reasons as residence of the souls of ancestors, as sources of medicinal plants or as natural water springs.[28] Not only are such areas a source of non-timber forests products (NTFPs) useful especially to rural communities and also to the wider society but they also perform important ecological functions.[29]

The bio-cultural diversity richness of tropical areas is of great interest to industries and research institutions for its commercial and scientific research potentials.[30] While temperate areas can also provide useful substances for the pharmaceutical and cosmetic industries, the attractiveness of tropical areas is much greater based as it is on the very high diversity of such research materials.[31] By implication, the conservation and long-term sustainability of the biodiversity components will ensure that tropical areas continue to be of interest to life-science researchers and bio-prospectors and to benefit future consumers of products arising from their investigations, which may include life-saving drugs.[32]

One study on the use of natural compounds isolated from plant species estimated that no fewer than 119 compounds derived from 90 plant species were important drugs used in one or several countries.[33] Furthermore, 77 per cent of these compounds were isolated from higher plants used in traditional medicine.[34] An example of a natural compound-derived drug is quinine. Quinine, which has so far has been the most successful drug for the treatment of malaria, derives from the bark of trees belonging to the genus *Cinchona* that grows naturally in the Andean region. For millennia, traditional populations in South America have used the bark of Cinchona in the treatment of fever.[35] Other examples are the drugs vincristine and vinblastine, derived from the rosy periwinkle (*Catharanthus roseus*), a plant originating from Madagascar. Eli Lilly, a US pharmaceutical company, and the University of Western Ontario studied this plant's use as a treatment for diabetes in the Philippines and Jamaica and isolated anti-cancer active substances rather than anti-diabetic ones.[36] However, despite such reported cases of modern exploitation of biological, genetic resources and TK, the full potential of biodiversity is not yet understood. It is estimated that less than 10 per cent of the many thousands of higher plants of use to humanity and on which traditional communities possess profound knowledge have been screened.[37] The current state of research suggests that there is an estimated 1.75 million species on earth, although other estimates suggest that the reality may be many times this number.[38] Other studies suggest that it will be hard for the modern screening techniques of phytochemistry to identify all therapeutic compounds inside the 80,000 plant species of the Amazon through random techniques that are not based on observation of local usage.[39]

Industrial research and development based on PGRs and associated TK acquired from developing countries is predominantly carried out in Western countries. While arguably research and development in this area may generate wealth to promote conservation and sustainable development in the source countries too, in practice this is not easy.[40] One of the main reasons is that developing countries lack the capacities to undertake profitable research and development based on their resources and are thus reduced to the role of suppliers of raw materials. The separation between centres of biological and cultural diversity richness and centres of research and commercial exploitation is therefore very obvious, as genetic resources move from the biodiversity rich south to the technologically rich north where they are transformed into valuable commodities.[41]

Natural product research and commercial value of genetic resources and TK: IPRs and contractual issues

The commercial value of genetic resources and associated TK is a highly debated issue, and using it as a justification for the establishment of workable regulatory measures for the protection of community rights is controversial. It should be pointed out that the exploitation of PGRs by corporate bodies (research institutions and commercial entities) in the areas discussed below is sometimes associated with the use of TK and practices of traditional societies who use the relevant PGRs in their livelihoods systems on a daily basis. Thus, arguably, such practices are generally captured by corporate users to guide subsequent research carried out in laboratories.[42] Several authors have reported the relationship between the use of drugs derived from pharmaceutical research and the traditional medicinal use of the plants having the main compound of such pharmaceutical drugs.[43] Considering that there are some exceptions to this argument,[44] and taking into consideration the evidences about the linkages between PGRs and their traditional uses discussed in the previous section, it rather seems logical to argue that the commercial value of PGRs should not be dissociated from the commercial value of TK associated with such PGRs.

Exploitation of biological resources generates considerable revenues to business operators. In the seed industry sector, when in 1994 some senior officials of the US Administration were pressurizing the US Senate to ratify the 1992 CBD, they used the economic contribution of foreign germplasm to the US seed industry to back their position. The then Secretary of State Warren Christopher pointed out that foreign germplasm added more than US$10 billion to the US$28 billion annual sales of the maize and soybean markets.[45]

The global trade in pharmaceutical and agricultural products is estimated at US $300 billion annually, of which the commercial sale of agricultural seeds amounts to US$30 billion.[46] Furthermore, the annual value of the botanical medicines industry is estimated at US$20 billion, while the sales in ornamental and horticultural products amount to US$16–US$19 billion annually. The cosmetic and personal-care industry produces annual sales estimated at US$50–US$75 billion annually. The average sales of pharmaceutical products made by the top companies is US$10 billion annually, between US$2 and US$10 billion in the cosmetic and personal-care products and US$200 millions in botanical products. In addition, the turnover of top seed companies, including Pioneer Hi-Bred, Monsanto and Novartis, is estimated at US$1 billion–US$2 billion annually with several small companies having turnovers of a few hundred million US dollars annually.[47]

The contribution of genetic resources to these figures varies from one sector to another. In general, natural products contribute around 20 to 50 per cent of the total sales of pharmaceuticals.[48] For example, the cholesterol-lowering drug Zocor, deriving from a fungi genus, contributed to Merck & Co.'s US$6.7 billion annual sales, representing 50 per cent of the company's total annual sales in 2001. Other studies addressed the possible links between natural products and drugs deriving from research and development. An analysis of data related to new drugs approved by the United States Food and Drug Administration (US FDA) or comparable entities in other countries from 1985 to 1995, focusing on cancer and infectious diseases, has been carried out.[49] The results of this study indicate that 62 per cent of the 87 approved cancer drugs were either of natural origin or were modelled on natural parents. Furthermore, 61 per cent of the 299 anti-cancer drugs that entered pre-clinical and clinical development were either of natural origin or were modelled from natural parents. A follow-up study published in 2003 analysed data covering an extended period of 22 years from 1981 to 2002 and assessed another sub-division of compounds related to natural products called the 'natural products mimic'.[50] This study indicates that, in relation to cancer, the percentage of small molecules, new chemical entities that are non-synthetic has remained at 62 per cent over the whole period. However, in relation to anti-hypertensive drugs, 48 of the 74 formally synthetic drugs can be traced to natural product structures/mimics, while among the 10 anti-migraine drugs, seven are based on serotonin. The importance of natural products as sources of new molecules, perhaps not necessarily as sources and new drugs, is therefore demonstrated.

However, to achieve the possible gains mentioned above, research institutions and industries undertake complex and complicated research and development processes which, in the context of the pharmaceutical industry, are time-consuming, risky and expensive, costing about US$400 million–US$800 million to produce a successful commercial drug.[51] In the USA, pharmaceutical research and development processes last between 10 and 15 years, at the end of which the FDA may or may not approve a particular drug for safety reasons.[52] The complexity of targeted diseases requires more time for study and, often, targeted diseases are difficult to evaluate.[53] In terms of risks, in about 5,000 to 10,000 compounds screened at the initial stages in the research and development processes, only 250 enter preclinical testing, five enter clinical testing and one is approved as a drug by the FDA. Only three out of every ten medicines brought to the market generate enough revenue to cover the average out-of-pocket cost of research and development.[54]

Considering that genetic resources and associated TK have direct and potential commercial value and are being exploited for scientific and commercial purposes, I reject arguments that tend to undermine their commercial or economic value. Assigning rights to rural communities over these biodiversity elements can therefore be justified on economic grounds. Such an approach reflects Demsetz and Lerch's economic justification of property-rights systems.[55] According to these scholars, if the economic value of certain resources increases, it is necessary to establish a system that defines the property rights associated with those resources because in the absence of property rights, the exploitation of such economically valuable resources will be difficult.

However, there is another approach that goes one step further to the one suggested by Demsetz and Lerch. Based on a proper allocation of rights over some economically valuable resources among parties involved in a relationship or contractual building process, Ronald Coase suggests that states promote private bargaining or contracts as the best practical approach to allow parties to appropriate the financial value of the resources to which their rights are related.[56] Contracts should, however, provide for some enforcement and other mechanisms to ensure that the use of the resources is sustainable. Some scholars discussed the Coasian contract in relation to the use of biodiversity as economically valuable genetic resources.[57] Sedjo and Simpson argue that by using the ordinary legal instrument of a contract, contractual arrangements specify the ownership rights of the genetic constituents of biodiversity to the country of origin of the species. Furthermore, in contractual agreements, what happens in effect is the fact that 'the source country trades the rights to collection in return for a guarantee of some portions of the revenues generated by commercial development of a product that utilizes a genetic constituent from a wild genetic resource collected within the country'.[58] However, the extent to which property rights over genetic resources can provide incentives to protect such resources on the basis of their financial value and the extent to which the revenues generated by a source country through a contractual agreement can stimulate biodiversity conservation are very unclear. As argued by Sedjo and Simpson 'current evidence indicate[s] that the financial values of genetic resources could be quite low by virtue of their redundancy (the presence of the same plant species in many locations) and thus limiting the potential of a property rights system to provide substantial conservation and protection incentives'.[59]

Taking the aforementioned arguments into consideration, it is my view that there should not be any doubt about the financial or economic values of community assets when exploited in scientific and commercially motivated undertakings. Indeed, PGRs and associated TK and practices of local communities have significant commercial value and are effectively being sought after and used by corporate bodies in their scientific and commercially oriented creative activities. These values are therefore strong arguments in support of any process or initiative aimed at establishing a property-rights system over such resources. However, whichever approach is the best in the sense of promoting sustainability (between a property-rights system and the contractual approach), rural communities must have the incentives and must be empowered to control access at local level. Ensuring full participation of rural communities in the decision-making processes as key partners is one of the strong motivations for their greater commitment to sustainability at local level. It therefore seems appropriate to clarify the grounds upon which traditional communities lay claims of ownership over PGRs, their TK, and innovation and practices associated with those resources. Below we turn our attention on this issue.

Basis for communities' claims of ownership over biological and genetic resources and TK

In traditional societies, the knowledge associated with some valuable cultural and biological assets generally hold profound spiritual meaning vis-á-vis these societies' livelihoods system. The words 'inalienable possessions' are used to portray the exclusive IP of indigenous communities over the knowledge which is passed to younger generations by elders during ritual ceremonies.[60] There are many terms used by programmes and advocates of community-based livelihoods issues to define their culturally rooted rights that are at the core of community livelihoods systems, each of them bearing its own merit. The Center for International and Environmental Law (CIEL), based in Washington, use the term 'Community Based Intellectual Property Rights' (CBIPRs), which CIEL conceptualized in 2000, within the framework of its Communities' Program. CIEL argues that the CBIPRs concept is purposely designed to be useful in advocating on behalf of local communities their rights to control and manage natural resources. CBIPRs are the rights held by local communities, in a communal manner, over the vital elements of their culture including traditional songs, crafts and sculptures, dances, the medicinal plants and the knowledge associated with these plants on their use at the local level (see the following section for a further elaboration of the elements upon which CBIPRs are rooted). All these elements are central in the livelihoods of local communities. Importantly, CBIPRs are not prone to commodification; they are held by the community, and any decision to sell or transfer these rights must be taken in a community context.[61] Let us briefly examine how the concept of community IP works in the context of specific traditional societies such as the populations of Pitjantjatjara in central Australia.[62] The most important songs in this community depict the history of their ancestors and the Dreamtime. These songs are divided into many categories and used in different ways. Some songs are used every day and this category is known as *inma nguntji*, meaning 'outside' or 'false' or 'false front songs'. However, another category, known as *inma mula* and meaning 'true' or 'inside songs' comprises songs performed during male ritual initiations only. Outsiders and women are not accepted in such ritual ceremonies during which young men face the powerful knowledge of ancestors. This mechanism ensures that, as inalienable possessions, the knowledge remains the exclusive IP of the group and will be only accessible to those who undergo the process of initiation. According to Weiner: 'all efforts are made to retain some part of possessions that mark out who the person is in relation to the past. A loss of this knowledge weakens the group identity and erodes its capacity to reproduce itself in the future.'[63] Findings from an extensive study carried out by anthropologist Tom Griffiths suggest that the sacred knowledge transmitted during ritual ceremonies helps the social group to determine itself in a historical dimension, linking the history of the group with its current identity.[64] Sacred knowledge is also highly valued by rural groups because it is related to mythic and real ancestors. It explains the social reproduction of the group, which is not limited to biological procreation but, importantly, is about the reproduction of personal identity.[65]

Anthropological studies suggest that the conservation and perpetuation of knowledge carried out by an individual within a specific group constitute the basis for belonging to the group and are the foundation for any claims about exclusive ownership.[66] Only those who can prove their active commitment to these axioms belong to the group that owns the knowledge and the related practices including but not limited to the traditional use of plants for their medicinal or cosmetic properties or traditional environmental knowledge applied in the realization of local environmental sustainability. For example, through a selective process based on utility, the Mbeere of Kenya have developed a wealth of environmental knowledge that they employ in the botanical classification of the local flora based on the utility of different species.[67] In a transgenerational manner, such knowledge is passed to younger generations through experimentation and ritual ceremonies and perpetuated within the community. It therefore appears that conservation and perpetuation of knowledge at the community level is inextricably linked to rural societies' systems and practices for the management of their bio-cultural assets. Their systems are essentially flexible as they promote free exchange of the resources among community members and are based on the basic principles of reciprocity and generosity.[68] In applying these principles in the management of their assets at the local level, rural communities are able to survive,[69] to generate more innovations and to conserve and use their resources sustainably.[70]

Community IPRs requiring legal protection

It is important to undertake a methodological and dynamic assessment of rural communities' claims of IP and resource rights prior to suggesting any practical regulatory measures to address their concerns. Non-community members may not be aware of how rural communities value and use their biological, genetic resources and TK on a daily basis or of the grounds for complaints by these communities that their traditional systems of resources management have been trespassed upon by outsiders in pursuit of scientifically motivated activities.

The IP and resource rights of rural communities are broad. They include rights to traditional arts and designs, music, traditional dance, stories and drama, TK of local minerals, traditional ecological knowledge pertaining to the conservation of natural landscapes and TK of the nutritional and cosmetic values of plants and other biomaterials.[71] According to Ekpere, the concept of community rights recognizes that the customary practices of local communities derive from a-priori duties and responsibilities to past and future generations of both human and other species.[72] Community rights therefore reflect a fundamental relationship with all life and are imbued with an innate demand for respect. However, this author acknowledges that this approach to and description of community rights is not accepted by the dominant Western world. Nonetheless, the purpose of recognizing and providing some form of legal protection to such rights is to recognize and protect the multicultural nature of the human species. In a given community setting, community rights over PGRs and TK are squarely linked with some forms of responsibilities and all govern the use, management and development of these assets. These rights and responsibilities are based on the communal ownership of the relevant assets which existed before the concepts of individual and private ownership and property emerged.[73] The issue of responsibility for the management and development of the community assets is very important at the community level. In respect of such responsibility, any dealing with the communally owned resources is subject to approval by the community, through the authority of its representative who may be its traditional leaders or the elected members of a locally established resource management group that may also be known as a common initiative group (CIG) or a community based organization (CBO). It must be stressed that, if the community representative has the decision-making power at the village level, any decisions arrived at by the said leader are hardly made unilaterally but rather mostly in a participatory manner. Upon grant of the authorization to access the community assets, users of such assets, mostly outsiders, are generally subject to the rules and restrictions embedded in cultural or religious customs.[74]

In this book, we indiscriminately use the terms 'biogenetic resources', 'biological resources' and 'biological materials' to describe the tangible assets of local communities to which their intangible assets, their TK and practices, are associated. Furthermore, and as hinted earlier, this book has chosen to use the terms 'traditional societies', 'local communities', and 'indigenous communities' indiscriminately, purposely to avoid dwelling upon the unnecessary debates regarding terminologies. Nonetheless, if I was asked to adopt a specific definition, I would be inclined to go for the definition of the term 'local communities' by the African Model Law, which is: 'human populations who live in distinct geographical areas. They generate, use, manage and pass on their biological wealth, knowledge, innovations and practices, which are governed by their own customary laws whether written or orally transmitted'.[75] The specific rights of rural communities for which this book investigates avenues for legal protection through workable regulatory measures include:

- their customary ownership rights over PGRs and land;
- their rights to fully participate in the decision-making processes regarding the management of their resources;
- their rights to monitor and control access to resources, to determine terms for access authorization, and to be consulted prior to any access and utilization of their resources;
- their right to fairly benefit from the utilization of the resources and TK; and
- their rights to stop any activity not respecting their traditional rules and to file lawsuit against anybody who infringes the terms under which access was allowed.

A number of declarations and statements of indigenous peoples' associations have been elaborated to promote the rights of rural communities in relation to modern exploitation of genetic resources and TK. In 2002, an assessment of some of these 'voluntary instruments' was carried out by Graham Dutfield, who identified about 15 such declarations and statements.[76] In his assessment, Dutfield finds that these instruments cover a wide range of issues, expressed as the critical demands from rural communities within the framework of access and utilization of their assets. Such critical demands of rural communities include but are not limited to their ownership and inalienable rights over knowledge and the resources, their PIC, the issue of participation, their rights to veto research or access to lands, knowledge and resources, the full disclosure of research results, demand for compensation and benefit sharing.[77]

There is no doubt that these declarations and statements bear significant legitimacy; but they do also have weaknesses. In terms of their strengths and legitimacy, they are frequently cited by scholars and policy analysts and are therefore well known and influential; they were agreed by representatives of many indigenous peoples' associations across the world. Consequently, they reflect some degree of international consensus. Moreover, these statements and declarations target a large international audience comprising a range of stakeholders having an interest in biodiversity regulatory policies including indigenous peoples, national and international policy-makers and intergovernmental organizations. In this respect, they are carefully drafted and very sophisticated. In respect of their limitations/weaknesses, one would argue that they offer no practical guidance for stakeholders to request the PIC of rural communities in the manner that recognizes them as the key partners in the decision-making processes pertaining to access. Also, they provide no practical guidance on how rural communities may veto access to the resources or lands on the basis that they have not been consulted beforehand. Overall, these declarations and statements have merely identified the above-mentioned critical demands without suggesting some practical measures to achieving them.

Taking into account the concept of community rights and other voluntary frameworks such as professional declarations, most of which are developed by advocacy bodies and aimed at promoting local communities' interests and expectations in line with their livelihood systems, it appears that the forms of protection achieved under the standard IP laws are not appropriate for the protection of community IP and resource rights.[78] The market-driven and profit-making features of modern IP concepts cannot be properly absorbed by local communities' systems of resource ownership and management, which operate through a system of local governance that is communal, participatory and inclusive, not private and exclusive. The 1967 Convention Establishing the World Intellectual Property Organization (WIPO), defines IPRs to incorporate rights in the creations of the mind, such as inventions, industrial designs, literary and artistic works, symbols, names and designs as well as all other rights resulting from intellectual activity in the industrial, scientific, literary or artistic fields.[79] The concept of IP is very broad and may be relevant to elements that are not currently covered by the existing types of IP, provided they result from intellectual activity in the industrial, scientific, literary or artistic fields.[80]

The owner of any modern IP is an identifiable individual or a legal entity who must have described his or her creation in the manner that reflects predesigned requirements. For example, the owner of patent rights on an invention based on scientific exploitation of genetic resources is an identifiable individual or corporation that must have transformed these resources and presented his or her invention in the manner that complies with the basic requirements for patentability notably novelty, inventiveness and industrial applicability. In this framework, the raw resources are transformed into commodities (the process of commodified and) henceforth owned by others. What other commentators have identified as 'commodification'[81] of knowledge characterizes the modern forms of IP and explains the market-driven aspect of these individualistic and monopolistic concepts.[82]

This book addresses the legal protection of community rights at the national level as an expression of countries' rights to develop national regulatory instruments covering issues of national interests. However, it is important that national policy and legal processes be aligned with any relevant international treaty or policy process of relevance to such issues of national interests. One such issue of interest, for which tailored and workable regulatory measures aimed at protecting the rights of local and traditional communities are needed in the countries rich in bio-cultural diversity, is the issue of 'bio-piracy' to which we turn our attention below.

The concept of 'bio-piracy' and its determinants

'Bio-piracy' refers to the use of genetic resources and traditional knowledge without the consent of rural communities and without providing them with fair compensation. According to Mgbeoji, there are three ways in which bio-piracy operates. The first is socio-cultural and relates to the constant denigration and denial of the intellectual input of traditional farmers and breeders, particularly women, in the nurturing of plants and the creation of TK of use of plants; the second is institutional and relates to the systematic collection of plants materials from the South and the setting up of an international system of gene banks; and the third is legal and relates to the patenting of plants and the TK of the use of plants.[83] This way of fractioning the concept of bio-piracy undoubtedly explains the extent of this concept and provides broader tools to all actors in their struggle to analyse cases that may be considered as 'bio-piracy' cases. Indeed, key in understanding the bio-piracy concept and to deciding whether a given case amounts to bio-piracy are indicators such as accessing the components of the assets of local communities, acquiring them and exploiting them in scientific and commercially oriented activities without the communities' consent and the application of modern IP to these assets or to their derivatives and benefiting from exploiting such IP without sharing with the source of the raw assets (local communities). It must be stressed that all of these activities are sometimes undertaken in disrespect to the legally established biodiversity/ABS regulations (where they exist). However, another look at the first component of the bio-piracy concept as described by Mgbeoji suggests to me that analyses of bio-piracy cases should not just limit themselves to associating this concept with illegal access to the assets of rural communities pursuant to the biodiversity regulations in force in a given country. From my perspective, and this is the standpoint of this book, whether ABS legislations exist or not and whether existing forestry regulations incorporate some ABS provisions or not, the simple demise of the socio-cultural contribution of local and indigenous communities in developing knowledge and sustaining their assets amounts to bio-piracy. It must however be stressed that the concept of bio-piracy is not evenly accepted by policy analysts, international development bodies and academic researchers, some actors still arguing that the various claims and complaints of advocacy groups sympathetic to the concerns of local communities over issues of bio-piracy amount to a limited understanding of this concept, which in turn make it vague and difficult to comprehend.[84] I refute these suggestions, sticking to my views that, indeed, the way we describe the concept of bio-piracy does not complicate its understanding. Therefore the bio-piracy concept is not just a matter of illegal access, acquisition and exploitation of the assets of local communities but also encapsulates a strong moral and ethical dimension, which is entrenched on the demise of the cultural linkages between the communities and their assets, from which local communities have contributed in nurturing and sustaining their assets. Based on our subscription and presentation of the bio-piracy concept, 'bio-piracy' is not simply a contemporary practice, so as to be merely viewed and accepted by some analysts in respect of failure by scientific and commercial actors to abide by current laws and professional codes of conducts, but rather is very much an old practice, which, in the context of the Amazonian region, is reported to have started as far back as in the 1500s. Be it today or back in the history, scientific and commercial actors who have shown a desire in acquiring the resources (for example, medicinal plants or plants with cosmetic properties or traditional germplasm) of indigenous and local communities and associated TK, have done so largely in pursuit of their specific interests. However, they have not always been very open to discuss their interests with their counterparts (local communities as the providers of the resources) so that they can act in an informed manner. In fact, in many respects there was no platform or room for discussion because the resources could have been taken by scientists from the colonial power on the back of the relationship between the colonial powers and developing countries. The historical dimension of bio-piracy which was caused by the complete lack of ethics and respect from the part of those acquiring the assets of the peoples is thus a very critical determinant of bio-piracy and should not be left aside in today's 'bio-piracy' discourse.

Put simply, to make a determination as to whether a scientific or commercial activity based on the assets of local communities amounts to bio-piracy, it is critical to assess some basic factors, which we call the bio-piracy determinants. One of the factors needing critical appraisal is whether the operator exploiting the assets acquired them according to a given country's legal procedure pertaining to access and use of biological resources and associated TK and practices of local communities. It is also important to establish if the operator conducted proper negotiations with the communities, disclosed its scientific and commercial interests and acquired the communities' PIC based on such disclosures and whether or not the operator signed any benefit-sharing agreement with the communities concerned. In addition, important aspects of the commercialization of research outputs resulting from the exploitation of community assets include the filing and acquisition of patents and PBRs. It is essential to establish whether the operator made any step towards commercialization and whether or not the operator is in partnership with an industry for the further commercial development of their research outputs. Not least, it is critical to establish the kind of benefits earned by the operator by their scientific or commercial activities and the extent to which such benefits have been shared with the providers of the raw assets, noting that such benefits need not be necessarily financial benefits. Failure from the part of a scientific and commercial actor to have addressed some of these issues should be enough to signify that its activities are squarely within the frames of the bio-piracy concept.

The cases of bio-piracy selected for each study country and discussed in Chapter 5 all have, in varying ways, the features and determinants of bio-piracy introduced here. Initially, it was the author's intention to have, among these cases, a balance of patents and PBRs-related bio-piracy cases. For the sake of convenience and simplicity, accessibility and availability to cases, the bio-piracy cases discussed in the book do not include PBRs-related cases. For one thing, it has been difficult to identify obvious PBRs cases in the study countries that harness the determinants described here. Part of this difficulty stems from the fact that, unlike traditionally used medicinal plants which can be traced back to a specific country and community in a fairly easy manner, the breeding materials used by seed companies and agricultural research institutions are almost never sourced directly from local communities. By the time plant materials get into the hands of seed companies and agricultural research organizations and are submitted to further improvements that may warrant legal protection through PBRs, they usually have already gone through many stages of selection and have been exchanged among professional plant breeders to the extent that it becomes very difficult for a given community to lay claim of traditional ownership over such materials. To the contrary, in respect of patent-related bio-piracy cases, it is generally easier to demonstrate the point and legal and ethical conditions of access to traditionally used medicinal plants and the associated knowledge of local communities by bio-prospectors and academic research institutions. Our focus to the cases analysed in the book for each country was therefore primarily meant to advance our arguments on the determination of bio-piracy cases based on the factors and determinants described herein.

Outline

The book comprises nine standalone chapters which are however coherently organized to compliment each other, making a comprehensive piece with a core message in the form of the suggested regulatory tools discussed herein. Chapter 1 has introduced the main themes to be discussed in the book and has presented its focus, objectives and goals. Chapter 2 elaborates on the fundamentals and key drivers of IPRs regimes, stressing patents and plant variety breeders' rights systems. Chapter 2 also discusses the concept of community rights as these rights are captured in selected international forums such as WIPO, CBD and the FAO IT-PGRFA, where issues of access to community assets are debated. Using some of the community-rights-related concerns identified earlier in the first chapter, Chapter 3 explores some national approaches that have attempted to protect these rights. Thereafter and in a logical trend, Chapter 4 deals with human-rights instruments and their coverage of the community-rights-related concerns. Discussion in Chapter 4 is rooted on some international and regional frameworks on human rights. Subsequently, Chapter 5 deals with the concept of bio-piracy in a relatively elaborate manner. It identifies and defines the determinants of bio-piracy and discusses specific cases per study country (Brazil, India, Peru and South Africa) stressing the extent and manner in which they impact on some selected community rights. While acknowledging that the application of PBRs over community landraces in a way which reflects the bio-piracy determinants will have the same effect on community-rights-like patents, this discussion has not covered a specific case on PBRs over farmers' traditional landraces. For the sake of convenience and simplicity, accessibility and availability to cases, the chapter has only covered patent-related cases. It seemed logical to us that the chapter on bio-piracy cases be followed by an elaboration on concrete soft (professional) codes of conducts and regional binding regulations addressing community-rights-related concerns. This is the thrust of Chapter 6, which is immediately followed by Chapter 7, which covers, in a broader manner, the suggested regulatory measures of the book. These measures are broadly and briefly introduced in Chapter 7 and discussed in more details and from a practical angle in Chapter 8. The measures suggested here are IP-based and human-rights oriented. As concluded in the final remarks – Chapter 9 – the implementation of these measures creates a platform for interaction and consultation among stakeholders involved in access operations. Consultations will enable the understanding by all actors of each of the rights systems (community rights on the one hand and IPRs on the other) discussed here, thus creating conditions and paving the way for the improvement of trust among stakeholders. A good grasp by local and indigenous communities of the drivers and raison d'être of IPRs regimes and of the operations of scientific and commercial users of their assets may mitigate their perceptions on bio-piracy. On the other hand, in the course of their operations, scientific and commercial users of community assets must accommodate the necessity to respect community rights from a human-rights perspective, knowing that failure to do so will raise concerns such as bio-piracy thus strengthening the distrust in their operations. Key problems with these measures lie with their international enforceability and wider applicability in the form of an international framework.

2 Patents, PBRs and community rights in international forums

As the main IP tools discussed in this book, this chapter seeks to introduce the key features of the patent and PBRs systems, focusing on the foundational basis of these systems and how they are applied and contribute to the appropriation of the assets of traditional communities and affect their rights. In addition, a brief exploration of the ongoing processes in some key international forums is carried out, stressing the extent to which community rights are taken into consideration in these processes and whether or not it is sensible to say that both IP-related and community-rights-related interests are dealt with in an equal manner in the relevant debates.

Patents and PBRs in TRIPS and UPOV: criteria, objectives and usefulness

Before dwelling on the main features of a patent system and its application to the protection of the various interests arising from research results based on scientific and technological exploitation of the assets of local and indigenous communities, let us first briefly indicate how the patent system emerged and spread.[1] The patent system originated from Florence, Italy in 1421, when the architect Filippo Brunelleschi, who became famous for designing and constructing the Santa Maria del Fiore Cathedral, realized that he needed some form of protection to the blocks of marbles up the river Arno from Pisa and decided to design a boat. But before this, he demanded and obtained something unprecedented, the sole rights given to him by the city of Florence over the design of the craft. In its very first trip, the boat sank with a large number of Filippo's belongings. But the acquisition of the sole rights over his design gave birth to the patent system, which spread into other European countries.[2]

Nowadays, users of PGRs and TK for scientific and commercial purposes increasingly apply the system to protect the products or processes derived from their inventive activities. But claiming patent rights on the products or processes derived from the exploitation of PGRs and TK raises a number of concerns to biodiversity-rich countries such as the lack of compensation when scientific and commercial activities turn out to be profitable to modern users. Moreover, the way the patent system operates seems to have, at least from the perspective of proponents of community rights, significant negative impacts over such rights. Likewise, there are complaints about misappropriation and 'bio-piracy'. But the question that needs to be asked is whether or not such concerns are grounded on a good understanding of the true extent and limits of patent rights?

The 1994 WTO-TRIPS[3] Agreement sets the minimum standards for IP protection including patents. Article 27(1) of TRIPS stipulates that: 'Patents shall be available for any inventions, whether products or processes, in all fields of technology, provided that they are new, involve an inventive step and are capable of industrial application.' A practical approach to describing a patent is to consider it as a scientific as well as legal document. A patent is a scientific document because it contains a full description of the underlying technology, enabling those persons skilled in the art to make and use the invention. In this respect, insufficient technical disclosure is a ground for disqualifying a patent application. A patent is also a legal title to a specified IP territory within which nobody may trespass in a bid to make any use without seeking a prior permission from the title holder. The exclusive rights are held by the title holder over a period of twenty years, such exclusivity being kept through payment of renewal fees over the course of the 20-year period. As stipulated in Paragraph 4 of Article 65 and Paragraphs 3 and 8 of Article 70 of the TRIPS Agreement, patents shall be available and patent rights enjoyable without discrimination as to the place of invention, the field of technology and whether products are imported or locally produced. These provisions represent a solid legal ground for those investing in the transformation of biogenetic resources and associated TK in the search for patent rights over the relevant outputs of their activities.

Novelty, inventive step (or non-obviousness) and industrial applicability are the three basic and key requirements under national patent laws[4] across the world and in regional agreements.[5] However, despite the growing harmonization of approaches in patent procedures at regional and international levels, patent rights should be analysed in the national context as: 'legal monopolistic rights lasting during a specified period of time and granted by States to the authors of inventions having potential commercial applications'.[6]

Many studies over the years have examined the objectives of the patent system.[7] Four key objectives emerge from this body of literature, notably: (1) increasing inventive activities; (2) disseminating knowledge about inventions;[8] (3) regulating the orderly investigation of new research areas; and (4) facilitating the practical use, including the production, application and commercialization of invention. At this juncture, it is worth assessing the economic role of patents in relation to the exploitation of PGRs and TK.

Patents are designed to create markets for knowledge by assigning property rights to inventors, enabling them to overcome the problem of non-excludability while at the same time contributing to maximum diffusion of knowledge into the public domain.[9] Kushan identifies two major beneficial effects: (1) the encouragement of investment in research and development needed to turn invention into marketable products and, (2) the fact that partners will exploit the market value of an invention in the way that would not have been possible without patents.[10] On these fronts, while rewarding inventors who made public their knowledge, patents promote research and development. Patent rights can be shared among partners, or the rights will be vested with one partner who will be expected to exercise them on behalf of the other collaborators on the basis of an agreement that secures everyone a share in the benefits derived from working the patent. However, in practice, if the filing of patents on inventions based on the exploitation of the assets of local and indigenous communities happens to yield substantial economic benefits, such benefits are likely to be enjoyed by groups other than their communities who live in the areas where these raw materials are collected.

Practically speaking, as long as the author of a particular invention is capable of describing his or her products or processes according to the three basic requirements for patentability, he or she is likely to be granted patent rights.[11] However, although it is possible in many countries to acquire a patent over scientifically proven new uses of old products including natural products, traditional societies are unable to file patents over their assets in their raw nature.[12] In this respect, it is hard to figure out how easily traditional and local communities would appreciate the usefulness of the objectives of patents to them.

PBRs, often considered to be 'patent-like' rights, are a category of IP protection accorded to those involved in the development of new plant varieties which offer imporoved features and qualities useful in agricultural and farming practices. These improved and useful features include, but are not limited to, varieties' resistance to pest and diseases, resistance to extreme and changing environmental conditions such as drought and flooding, and other qualities such as taste and improved nutritional value. There are many similarities between patents and PBRs, but there are also obviously significant differences between these two systems, explaining why PBRs are called 'patent-like' rights.[13]

As a category of IP, the protection of PBRs follows – as the case with patents – two broad foundational approaches. The first approach is the one which is based on the view that the development of new plant varieties is a process that is engraved with the personality of the developer. In this respect, it is very simply just morally correct to assign to the developer the right of ownership of the new plant variety, thus legally providing him or her with the opportunity of making some economic gains from this product of their ingenuity. This moral claim of ownership of the product of mind (the new plant variety) is consistent with the pursuit of human-rights protection, in line with Article 27 of the Universal Declaration of Human Rights, which provides for the protection of: 'the right to the protection of moral and material interests resulting from scientific, literary or artistic production of which he is the author'.[14] Based on this dual moral and economic premise, the legal protection accorded through the PBRs scheme is meant to reward the breeder for their intellectual input and creative mind, as well as for the time and financial investments they invested in the development of the new plant variety.

The second justifying argument for the protection of PBRs reflects an instrumentalist approach to the protection of IP in general. Based on the instrumentalist view to the protection of IP, PBRs are not justified on the grounds of morality, or economic or financial reward that they provide to the developer of a new plant variety. Rather, the protection of PBRs is justified primarily on the basis of the contribution plant breeders make in bringing more products to society and in the enhancement of societal knowledge and welfare.[15] This instrumentalist approach to the protection of IP is clearly consistent with the objectives and practical usefulness of patents as discussed earlier. From an instrumentalist perspective, a PVP system that does not succeed in addressing society's needs in new agricultural materials will not be pursued, let alone achieve its objectives and practical purpose effectively.

Legally speaking, PVP systems are internationally crafted in two major forums or conventions, the UPOV and the WTO-TRIPS Agreement.[16] From its original version, which was adopted in 1961 and entered into force in 1968 to its most recent version (the 1991 revised convention), UPOV provides the only internationally recognized *sui generis* system for the protection of new plant varieties. The convention has been revised three times, with the effect of more and more reduction of the privileges of farmers and undermining of farmers' rights, while witnessing an increased broadening in the scope of protection, expansion and strengthening of PBRs. In particular, the 1991 version of UPOV has had the effect of diminishing the rights of farmers to reuse seeds saved from their previous harvest which might have derived from protected varieties, while, at the same, it has expanded the rights of plant breeders over what is technically known as essentially derived varieties (EDV).[17]

Like patents, there are basic criteria that plant varieties need to comply with to enable their improver or developer to be granted exclusive rights over them. For plant varieties to be eligible to protection as provided under the UPOV framework, they must be new,[18] distinct,[19] homogeneous (or uniform)[20] and they must be stable.[21] If these criteria are very relevant to the conditions under which plant breeders operate and the methods they apply in the development of new plant varieties, they are criticized with regards to their impacts on community rights, community-based farming practices and agricultural diversity. From the perspective of farmers, plant breeders use their traditional landraces to produce the new plant varieties that are consistent with the provisions of the UPOV convention. Once they are granted PBRs on improved varieties, farmers immediately fall under the threat of being blamed of piracy if they exploit commercially their traditional landraces which are similar to the protected variety for which plant breeders hold PBRs. Here, one can see the strength of the distinctness criterion as it stretches the scope of protection of PBRs while seriously undermining farmers' ownership rights over their traditional landraces. Also, the criteria of homogeneity and stability are criticized in respect of their damaging effect on agricultural diversity which is at the heart of traditional farming practices and thus food security in farming communities. Indeed, for a plant variety to fulfil the homogeneity criterion, it has to be: 'sufficiently homogenous, having regard to the particular features of its sexual reproduction or vegetative reproduction'.[22] The UPOV guidelines further clarify that the homogeneity criterion is fulfilled if the 'variation is as limited as necessary to permit accurate description and assessment of distinctness and to ensure stability'.[23] However, traditional farming practices utilize traditional landraces which are much diversified. By virtue of the extent of such diversification and variations embedded in traditional landraces, farming communities have a wide spectrum of choice of materials best adapted to their farming circumstances. Using these varieties enables them to minimize the risk of losing their crops in the event of any attacks by pest or disease to their farms, which homogeneity and uniformity criteria and the stability criterion will not be able to prevent. But without demonstrating compliance to the homogeneity criterion, for instance, farmers are unable to secure PBRs over their landraces upon which slight improvements are enough to ensure their eligibility for protection through PBRs systems with exclusive rights granted to the improver. This assumption is, however, subject to debate because, as some analysts have argued, new plant varieties developed by plant breeders are not necessarily directly based on the traditional landraces of farmers but rather are accessed from gene banks, which generally host materials that have already been improved to a certain extent and are likely to blend properties and features deriving from materials sourced from several geographical areas.[24]

In strengthening the internationalization of PVP systems, the WTO-TRIPS Agreement requires member countries to provide for some form of legal protection to plant varieties through domestic law. However, TRIPS leaves it up to the discretion of countries as to what form of protection system they want to set up at the national level. Indeed, Article 27(3)(b) of TRIPS provides that 'Members shall provide for the protection of plant varieties either by patents or by an effective *sui generis* system or by a combination thereof.' There are varying interpretations of what TRIPS mean by 'effective' *sui generis* system. From the perspective of the plant-breeding industry, the UPOV framework provides for the effective *sui generis* system contemplated by TRIPS. This is the view that is advocated by developed countries' governments, which, by means of bilateral trade agreements, they coerce developing countries to sign up to UPOV as a means to fulfil their obligations under TRIPS. But developing countries, hosting large farming populations whose seeds' needs cannot necessarily be satisfied through their countries adhesion to UPOV, contend that TRIPS does not specifically mandate UPOV as the 'effective' *sui generis* system. From this angle, they are free to shape systems of PVP that reflect the realities of their national socio-economic and cultural circumstances and national priorities.

As seen, the international frameworks regulating patents and PBRs through TRIPS and UPOV encapsulate the core requirements agreed by the international community. But the manner in which key aspects of these systems are interpreted and implemented by technologically advanced countries which are better equipped to use these frameworks and to use them to their advantage, indicate that national approaches in respect of protection of IP is a matter of national priority. In the following section, we explore how one key feature of these systems, the 'prior art', is differently interpreted and applied in the determination of novelty across jurisdictions and what impact such varying interpretations have on PGRs and associated TK originating from developing countries, bearing in mind these countries' poor technological capacities to exploit their own assets.

Prior art and public domain knowledge in the USA, Japan and Europe

Before granting a patent, patent examiners must undertake a search and examination of the patent application. In theory, the three basic requirements must be thoroughly examined during these exercises. The search and examination procedures are meant to assess some of the features of the patent application including a description and the claims of the invention that can be presented in the form of drawings, diagrams or plans.[25] An important aspect of the description of the invention is the full disclosure of the invention to the patent office. In view to comply with the full disclosure requirement, the patent applicant is expected to present both his/her invention and the information already known to the public[26] as well as the critical steps which led to the said invention.

During the examination process, the examiners must ensure that the product or process for which the legal protection is sought is actually 'new' and derives from an inventive step,[27] meaning that it 'could not have been deduced by a person of average knowledge of the relevant technical field'.[28] If patent examiners conform strictly to these requirements, it is expected that they grant patents only over the inventions for which the information disclosed in the patent application has demonstrated a clear demarcation from what was already publicly known, which is recognized as 'prior art'.

Interestingly, patent laws provide that anything in the public domain, meaning that it is part of 'prior art', is not novel, and thus cannot be patented. TK is essentially transmitted from one generation to another and, generally, is commonly shared by a cross-section of traditional communities. In this respect, perhaps, it typically characterizes the 'prior art' concept. In its raw form, this knowledge cannot be patented or privately owned by individuals or corporate bodies, especially considering that the primary holders of this knowledge cannot, both on legal and practical grounds, file patent applications.[29] However, as discussed below, while in theory there is consensus among the major patent systems about the unpatentability of what is already largely known by or used in the public, there are varying interpretations of the state of the art concept from one regime to another, making it practically possible for something to be patentable and patented in one jurisdiction and not in the other.

In relation to the examination of patent applications, in some jurisdictions examiners limit the search for 'prior art' to their national or regional borders. This implies that the knowledge – such as the TK of folk medicines and the information contained in an expired patent – held in other territories might not be accessible to patent examiners in a foreign country or admissible in a court. Such a geographical limitation approach may facilitate the granting of patents that should not have been granted. As far back as in 1836, the geographical limitation of prior art in a patent application appeared in the US Patent Act 35 USC (United States Code).[30] Section 102(b) of the US Patent Act provides that:

> a person shall be entitled to a patent unless the invention was patented or described in a printed publication in this or a foreign country or in public use

or on sale in this country, more than one year prior to the date of application for patent in the United States.

Following this provision, knowledge held in a foreign country can be used to challenge a patent in the USA on the grounds that information disclosed within the realm of 'prior art' was published or included in a previous patent at least one year before the date of application in the USA.[31] Otherwise, knowledge in the public domain in a foreign country will legally fall outside the scope of the public domain according to the US system, and, consequently, the novelty requirement will be deemed to have been satisfied.[32]

In the case of Japan, inventions that are industrially applicable can be patented. However, they might not be if they were known orally, published or already in use in one way or another within or out of the country.[33] At first glance, the Japanese approach to 'prior art'[34] looks different from the geographical limitation approach of the US system. However, despite its rather blurred approach, the guidelines to the patent examiners in the Japan patent system include the knowledge already available in a publication, making it difficult to challenge a patent covering an invention that is supposedly based on such knowledge. To be able to challenge the novelty requirement of an invention, the knowledge described in a publication must be 'enabling'; that is, it must be sufficient to assist a person not skilled in the relevant domain to undertake the claimed invention.[35] It is only if this knowledge satisfies the 'enabling' criterion that, 'by law' in Japan, the invention may be denied patent protection on the grounds of lack of novelty.

With the rapid pace of development in biotechnological research, it will obviously be hard to demonstrate the extent to which TK has contributed critically to a given invention. The prior art related to an invention based on TK is normally assessed in respect of modern technology and the knowledge of industrialized countries. Patents are filed over products and on new and inventive processes.[36] Therefore, although a certain community may be able to prove that it has maintained and used its knowledge from time immemorial, such knowledge and associated practices may not be sufficiently enabling, as defined under the guidelines in Japan, to be legally considered 'prior art'.

In Europe, Article 54 of the Convention on the Grant of European Patent (EPC)[37] provides that for an invention to be patentable, it should not be part of the state of the art, which is considered to 'include everything made available to the public by means of a written or oral description, by use or in any other way before the date of filing the application'. There is no geographical limitation in this approach; neither is there any supplementary condition such as the 'enabling' requirement of the Japanese guidelines to patent examiners. The difference in the treatment of foreign knowledge between the US and European systems is evidenced on the commercial exploitation of the 'neem tree' and the associated knowledge of Indian communities and the patenting of related products both in the USA and Europe.[38] While the United States Patent and Trademark Office (USPTO) refused to revoke any of the neem-tree-related patents,[39] on the request of the 'Neem Team' composed of two Indian non-governmental organizations and the Belgian

Minister of Health and Environment, the European Patent Office (EPO) revoked patent EPO No. 0436–257 on 10 May 2000 and upheld its decision on 8 March 2005. Arguably, if the search and examination had covered foreign knowledge more effectively, perhaps the EPO patent on neem would not have been granted in the first place.

Slightly modified biological matter can therefore be patented, allowing private and corporate actors to own the resulting legal monopolistic rights, while rural communities are not able to own such legal rights over their raw assets. Indeed, to the extent that TK associated with biogenetic resources may be used in scientific research, depending on the jurisdiction, there may or may not be a possibility for a challenge to the patentability of the research outputs. Some commentators have argued that searchable databases of TK and practices of local and indigenous communities, made available to patent offices for the purpose of a prior art search, may be useful in preventing the grant of some controversial patents associated with these assets.[40] At least theoretically, patents on inventions based on biogenetic resources and TK would not be granted if such inventions are obvious to traditional communities. Obviously, to achieve this, there needs to be changes in the patent procedures to enable patent examiners to accept evidences from searchable databases. But would there be any consequences if patent laws were reformulated to accommodate the protection of TK? If patent laws were modified to provide for the protection of TK, existing TK-related patents may prevent traditional communities from obtaining further patents on their knowledge, innovations and practices.[41] Moreover, despite the emergence of strong scepticism,[42] the searchable databases will prevent traditional communities themselves from obtaining patents over their TK and practices because they will effectively be part of the databases. The communal ownership of TK and local innovations and their geographical spread across communities and countries pose serious obstacles in the establishment of inventorship[43] and may even raise conflicts among communities fighting for such inventorship.

The patent system therefore appears to be essentially twisted to favour modern knowledge systems. Does this, however, suggest that patent rights undermine the customary systems of the use and management of resources operating in rural communities? And is there any harm to the communities? This is an area of obvious controversies which this book investigates subsequently in the context of its examination of the bio-piracy cases. However, our point of departure in this debate is that patent laws can hardly be blamed for not protecting TK or the genetic resources. When these elements are transformed using modern technological means that facilitate compliance with the requirements for patentability, the derivatives can be patented. This patent is not really a 'bad' patent in the sense of causing a particular harm to rural communities unless, for instance, it ends up preventing them from accessing and using their resources as they wish, such as in small-scale commercial activities or for exportation. However, in some instances, rural communities may argue that the modern transformation would not have been possible in the first place without their knowledge and practices associated with the genetic resources. In this respect, protecting the products or processes derived from such transformation by a patent, arguably, strengthens rural communities' perceptions that modern IPRs – especially patents – facilitate the piracy of their possessions and that free riding over their valuable assets is unfair and exploitative.

Community rights in the CBD processes: working groups on Article 8(j), ABS and protected areas

The mechanisms of the 1992 United Nations CBD through the activities of the Conference of the Parties, the specific working groups and groups of technical experts have not embraced the obligations of States vis-á-vis the realization of the human rights of indigenous and local communities in a strongly worded and practical manner that can mimic the seriousness of these CBD settings to make a human-rights approach central to its mechanisms. Rather, what seems to be happening within the framework of the CBD's operations is merely that mention is made of the Convention's recognition or awareness about human-rights-related policy processes happening in other forums which deal with broader human-rights concerns of indigenous peoples. This trend is obvious in the documents of the decisions of the Convention of Biological Diversity – Conference of the Parties (CBD-COP) starting from COP 3 held in Buenos Aires, Argentina from 4 to 15 November 1996. Making its first appearance in the Decisions of the CBD COP meetings, the term 'human rights' appears in one of the preambular paragraphs of Decision III/14 on the implementation of Article 8(j), which deals with the rights of indigenous peoples and local communities over their TK, innovations and practices. In this preamble, parties merely take note of relevant activities – on the rights of indigenous peoples and local communities – occurring within the framework of the system of the UN and other international forums such as Convention 169 of the International Labour Organization. The softness of the tone used by this decision in referring to human rights can be understood in this preamble, as effectively the preamble is not a key treaty area where strong obligations and strategic actions for the realization of such obligations and actions are really stated. However, further down in Paragraph 4 of this decision, the Executive Secretary of the CBD is requested to

> remain informed as to relevant international treaties and bodies including, *inter alia,* those under the auspices of the United Nations Commission on Human Rights and Commission on Sustainable Development, Convention 169 of the ILO, the World Bank, the FAO, the WIPO, the United Nation Educational, Social and Cultural Organization, and the WTO and to provide periodic reports related to Article 8(j) and related articles to the conference of the parties.
>
> (Decision III/14 on Implementation of Article 8(j), 1996)

It therefore looks as if, early on in its activities, the conference of the parties to the CBD did not take up a human-rights approach as a key element of its activities regarding the concerns of indigenous and local communities over their biogenetic resources and associated TK, innovations and practices. Making human rights a core aspect of its activities would have meant that parties to the CBD had agreed, for instance, to incorporate a strongly worded statement in Decision III/14 requesting the Executive Secretary to create a working group or group of technical experts on human rights and the implementation of Article 8(j). The CBD secretariat should have then sought to establish partnerships with any other human-rights-focused forums and to request the full participation of its experts in their policy processes while allowing experts from the other forums to get involved in the activities of the CBD COP regarding the implementation of Article 8(j) and Article 15 on ABS. These partnerships would then have provided the platform for the realization of synergies among the various forums, making human-rights considerations central to the workings of the CBD and making biodiversity issues, TK, and the innovations and practices of indigenous and local communities be discussed specifically at meetings of the other forums. But this did not happen, and the trend of vagueness in respect of the inclusion of 'human-rights' language in the CBD COP decisions proceeded through the subsequent meetings of the COP. However, there has been a significant improvement in recent years, in respect of the collaboration between the CBD secretariat and the United Nations Permanent Forum on Indigenous Issues (UNPFII). This collaboration was materialized by the publication in the decisions of the Ninth Conference of the Parties (COP 9), held in Bonn, Germany, from 19 to 30 May 2008 of the Recommendations of the Permanent Forum on Indigenous Issues under Decision IX/13 on Article 8(j) and Related Provisions. Despite the slow progress of CBD COP proceedings in addressing indigenous peoples' and local communities' concerns as human-rights, such slowness that could be approximated to a disregard by the CBD COP – and related working groups – of indigenous peoples' human-rights, the inclusion of these recommendations in Decision IX/13 is nonetheless evidence of the strong ties betwen the CBD and UNFII. One of the key contributions of UNPFII to the work of the CBD on indigenous peoples issues included in these recommendations was the organization of an International Expert Group Meeting on the CBD's IR on ABS and Indigenous Human Rights. This meeting was held in New York from 17 to 19 January 2007 and yielded a report on indigenous TK prepared by the Secretariat of the UNPFII.[44] One can only hope that with the adoption of the UN Declaration on the Rights of Indigenous Peoples, the UNPFII will get busier to seek its inclusion in other policy processes where the rights of indigenous peoples and local communities over their biological resources and associated TK are at stake. One of those being the ongoing negotiations of the IR on ABS.

Notwithstanding the above, the extent of consideration of the concerns of local communities within the framework of the policy processes pursued under the umbrella of the 1992 CBD can be assessed through, on the one hand, its provisions pertaining to community rights and, on the other, in respect of ways in which the international community, through the meetings of the Conference of the Parties and its working groups is contemplating the implementation of these provisions. Two sets of specific provisions are of particular interest, those being Article 8(j) addressing TK, innovations and practices of local and indigenous communities and Article 15 on ABS. It is worth bearing in mind that, countries – not local communities – are the parties to the CBD. They are the entities which are expected to exercise the sovereign rights over the resources found in the national territories, pursuant to Article 3 of the CBD. However, in exercising this right of sovereignty, countries can recognize the critical role that local and indigenous communities play in the sustainable management of biological and genetic resources based on the application of their TK, innovations and practices. Such recognition can therefore be crystallized in national legislations and policy measures, which can give effect to Article 8(j) and Article 15 at the national level, by empowering local and indigenous communities to control access to the resources found in their territories. Access to such resources should practically be based on the PIC of indigenous and local communities and provide them with a fair share of the benefits arising from the exploitation of their biogenetic resources and associated TK, innovations and practices. Before turning to such measures in the subsequent chapters of this book, let us first briefly assess how, for instance, community rights of PIC and participation have been taken into consideration in the policy processes pertaining to the CBD's approaches to the implementation of Article 8(j) and Article 15.

Decision IV/9 of the CBD Conference of the Parties established a working group on Article 8(j) at the fourth meeting of the Conference of the Parties held in Bratislava in 1998. This decision states that:

> an open-ended inter-sessional working group be established to address the implementation of Article 8(j) and related provisions of the convention. For the purpose of implementing Article 8(j), Decision IV/9 mandates the working group, among other issues to: (a) provide advice as a priority on the application and development of legal and other appropriate form of protection for TK, innovations and practices of local and indigenous communities embodying traditional lifestyles relevant for the conservation and sustainable use of biological diversity and; (b) provide the conference of the parties with advice relating to the implementation of Article 8(j) and related provisions, in particular on the development and implementation of a programme of work at the national and international levels.
>
> (Decision IV/9 on Implementation of Article 8(j), 1998)

Since then, the Programme of Work for the implementation of Article 8(j) was developed and adopted at the COP 5 in Nairobi under Decision V/16 on Article 8 (j) and Related Provisions. The programme of work includes a list of issues that parties to the CBD have committed to address in the realization of their commitment to the implementation of Article 8(j) and Related Provisions. Furthermore, at COP 7 of the CBD held in Kuala Lumpur, through Decision VII/16, parties agreed on the development of draft elements of, and to commit to the implementation of, a plan of action which encapsulates elements for the retention of TK, innovations and practices embodying traditional lifestyles relevant to the conservation and sustainable use of biological diversity.

In respect of the extent to which community rights are taken into account in these CBD processes pertaining to the implementation of Article 8(j), it can be said that there is a certain degree of consideration of community rights in these processes. For one thing, the issues retained in the programme of work and in the plan of action cover comprehensively the rights-related concerns of indigenous and local communities in respect of access to, use of and sustainable management of biological resources and associated TK, innovations and practices. Indeed, the tasks identified in the programme of work were done through a process based on such principles as the:

> [p]rinciple of full and effective participation of indigenous and local communities at all stages of the identification and implementation of the elements of the programme of work as well as the full and effective participation of women of indigenous and local communities in all activities of the programme of work.[45]

More specifically, Task 1, Task 2 and Task 7 of Phase 1 of the programme of work deal with issues very dear to indigenous and local communities, notably their prior informed approval or consent and full participation at all stages of decision making on issues affecting their TK, innovations and practices embodying their traditional lifestyles and the issues of equitable share of the benefits arising from the exploitation of their assets.[46] Apart from significantly taking indigenous and local communities rights-related concerns into account in the processes pertaining to the implementation of Article 8(j), the CBD involves indigenous and local communities, through their representatives, in the activities of some technical groups. As evidence of this, representatives of indigenous and local communities are members of the advisory expert group, which was established in response to Decision VII/16, Paragraph E. This advisory expert group is tasked to produce the composite report on the status and trends regarding TK relevant to biological diversity.[47] Also, or rather more importantly, in order to facilitate participation of indigenous and local communities to the meetings of the CBD, Decision VII/16, Paragraph G established a voluntary fund which will be used to support eligible representatives of indigenous groups to the meetings.[48]

It is one thing for the international community to agree to discuss the rights-related concerns of indigenous and local communities within the framework of international processes such as those occurring within the UN CBD, but it is quite another for the international community to take actions, for instance in the sense of (a) legislating on these issues in an effective manner and (b) actually empowering local and indigenous communities to express their concerns at the critical moments of the deliberations during the meetings of the working groups. From our perspective, this is where the CBD processes can be described as weak or as failing to portray seriousness about addressing community interests and expectations and promoting and protecting their rights. This failure can be briefly examined in two distinct setups: participation of local and indigenous communities in the deliberations of the Working Group on Protected Areas and the evolvement of the CBD processes aimed at the implementation of the provisions of Article 15 of the Convention on ABS.

The Ad-Hoc Open Ended Working Group on Protected Areas was established through Decision VII/28 of COP 7 to the CBD held in Kuala Lumpur in February 2004. The main role of this working group is to support the implementation of the programme of work on protected areas, also agreed by the CBD COP parties under the same decision. One of the identified (original) mandates of the working group is to:

> contribute to further development of 'tool kits' for the identification, designation, management, monitoring and evaluation of national and regional systems of protected areas, including ecological networks, ecological corridors, [and] buffer zones, with special regard to indigenous and local communities and stakeholder involvement and benefit sharing mechanisms.[49]

The mention of special regard to indigenous and local communities' involvement implies that, in the process of identification, planning and implementation of the activities associated with this mandate, there needs to be full participation of local and indigenous communities with their views taken into account at critical decision-making stages. This does unfortunately not seem to be the case as evidenced by the incident that occurred during the second meeting of the Working Group on Protected Areas, held in Rome from 11 to 15 February 2008. Indeed, during the proceedings of this second meeting of the Ad-Hoc Open Ended Working Group on Protected Areas, representatives of indigenous peoples, organized through the IIFB, complained that they were not given the opportunity to discuss issues of concerns to them in a timely and appropriate manner. This failure to allow them to speak out therefore goes a long way to weakening their right to participating in the CBD processes because the final communiqué will not reflect their views. With this realization, representatives of indigenous peoples withdrew from the meeting.[50] As indicated by these peoples, failure to realize their full and effective participation in these processes is in stark contradiction with the decisions of the CBD and trespasses their rights as enshrined in the 2007 UN-DRIP.

The evolvement of negotiations on the implementation of the ABS provisions of Article 15 of the CBD en route to the IR on ABS displays the lack of seriousness of the CBD processes over issues of interests to developing countries rich in biological and genetic resources and traditional communities. Arguably, and as the CBD secretariat might claim, progress is made in the right direction, that is towards the completion of the negotiations of the IR on ABS by the deadline, which is by the tenth meeting of the CBD COP scheduled in Nagoya, Japan from 18 to 29 October 2010. Nonetheless, these processes illustrate the divide between the interests of the rich and technologically endowed countries, which can best use the biological resources and TK, and those of developing, technologically poor but biodiversity-rich countries that host large populations of indigenous and local communities. Rather than going into the details of the CBD processes related to the implementation of the ABS provisions of Article 15 of the CBD, let us just map out the key stages of these processes and demonstrate the CBD failure to address community rights and concerns consistently. At the commencement of its work on the implementation of the ABS provisions of the CBD, parties to this Convention began with information-gathering. They decided to speed up work in this area at COP 5 in 2000, held in Nairobi, Kenya, by adopting Decision V/26 which created the Ad Hoc Open Ended Working Group on ABS. The initial mandate of the ABS working group was to develop guidelines and other measures on ABS and to help parties in their implementation of the ABS provisions of the CBD. A key achievement of this working group was the drafting of the Bonn Guidelines in 2001, which were eventually adopted through Decision VI/24 at COP 6 held in The Hague in 2002. However, despite the rather comprehensive elaboration of these guidelines on ABS and its guidance to countries wanting to embark in the development of national ABS measures, they are no more than just voluntary guidelines. In contrast, despite the reluctance or refusal so far by the USA to ratify the CBD, it is still the case that the CBD is a legally binding biodiversity conservation agreement in which parties managed in the course of the negotiations to incorporate strong provisions on biotechnology and patents in respect of biotechnological exploitation of the resources that are likely to be sourced from biodiversity-rich but technologically poor countries. As discussed earlier, patents are monopolistic rights that work best for countries that have the technology muscles to undertake state-of-the-art scientific research and transformation of biological resources. Not making the Bonn Guidelines legally binding has been, from my perspective, a failure by the CBD processes to properly address community-rights-related concerns in respect of, for instance, PIC and benefit sharing.

A similar trend is observed within the framework of the negotiations of the IR on ABS, which are currently underway under the aegis of the Ad-Hoc Open Ended Working Group on ABS. In 2006, the draft text of the IR, submitted by the Working Group on ABS and published in Decision VIII/4 at COP 8, was highly bracketed. These brackets were then seen as evidence of the strong disagreement between the developing and bio-cultural diversity-rich countries and the technology-rich but bio-cultural diversity-poor countries. Developing countries in particular felt that issues of major concern to them and to their indigenous and local communities are those that represent the sticking points of the negotiations and are thus the highly bracketed ones. Indeed, a quick scan through the draft text of the IR published in Decision VIII/4 shows that the rights-related issues of concern to developing countries including but not limited to PIC, disclosure of source or origin of the resources in patent applications and disclosure of evidence of benefit-sharing arrangements, were the areas that were mostly in parentheses. In addition, developing countries have consistently said that they expect the IR to be a legally binding instrument. But, this position is highly rebutted by the developed countries, which, in actual fact, are gearing these negotiations towards an output that may be a non-legally binding instrument or an output that may comprise many pieces of small instruments, which may equally be non-legally binding and the implementation of which may end up being cumbersome.

COP 9, held in Bonn Germany in 2008, praised the progress made by the Working Group on ABS in advancing works on the negotiations of the IR. This approach implies that Annex 1 of Decision IX/12, which henceforth represents the basis for further negotiations of the IR, was a significantly advanced version of the highly bracketed text of the IR annexed in Decision VIII/4 of COP 8 held two years earlier in Curitiba, Brazil. This is true, considering that the text resulting from COP 9 included less brackets than its earlier version. Also, parties had undoubtedly moved forward in making concrete proposals on the options to consider on the various key areas of the IR, notably on its nature, scope and objectives. Nonetheless, there are still major divergences of views on key issues and diverging positions in the use of terms, which demanded, yet again, the establishment of three groups of technical experts. Indeed, three such groups – on compliance; on concepts, terms, working definitions and sectoral approaches; and on TK associated with genetic resources – have been established.[51] A comprehensive deliberation of the outcomes of the meetings of all the groups scheduled to take place at the Ninth Meeting of the Working Group on ABS will hopefully set the scene for the conclusion of the negotiations of the IR before COP 10. It is, however, hard to see how developing countries' delegates will influence the direction of the proceedings in the technical groups of experts, while one may even ask what is the big deal in setting up yet other technical groups, whose outcomes rarely, if ever, reflect the views of developing countries and may hardly change or be influenced by outcomes from regional policy processes. For example, the African position on the definition of and the objectives and scope of the certificate of origin/source/legal compliance is not similar to the outcome of the working group of experts that was set up to address this issue within the framework of the negotiations of the IR. The working of experts that met in Lima, Peru, in January 2007 settled that the certificate that could ensure the traceability of genetic resources from the moment of access through its entire chain of utilization should be named the 'Certificate of Legal Compliance'. But during a workshop (which the author of this book personally attended) of African experts on ABS organized in Addis Ababa, Ethiopia, from 28 to 30 March 2007, an African position emerged which was to be channelled to the African Negotiators of the IR. The Addis Ababa position clearly stated that the certificate should be called 'a certificate of origin and source' or 'the certificate of origin and/or source'. This position from Africa has not changed the course of the negotiations in respect of the appellation of this certificate, because, as they stand, ongoing negotiations of the IR on ABS are based on the Lima position which remains as the certificate of legal compliance.

It is hard to anticipate the outcomes of any international negotiations. However, their outcomes generally reflect the views and positions of the most powerful countries, which are always very well represented by highly trained and experienced delegates. In the context of the negotiations of the IR, it is hard to see how they will result in a legally binding instrument and how issues such as the requests for disclosures of PIC and evidence of benefit-sharing arrangements in patent applications will feature prominently in the final text.

Community rights in the WIPO IGC processes

Established in October 2000 by the General Assembly of the WIPO, the Intergovernmental Committee (IGC) on TK Genetic Resources and Folklore is a forum for debates within the WIPO over the issues at the interface of IP, TK, genetic resources and expressions of folklore. A critical element of the mandate of the IGC, as defined by the WIPO General Assembly, is to consider the international dimension of questions related to the legal protection of TK, genetic resources and expressions of folklore, including and without prejudice to similar proceedings taking place in other forums and the development of an international instrument or instruments.

Since its inception, the meetings of the WIPO IGC, which are all held at WIPO headquarters in Geneva, have been characterized by strong divergence in the participating countries' perspectives about the legal protection of TK and expressions of folklore within the framework of IP systems. These disagreements cut across a range of issues which are very relevant to developing countries, where most of the biological and cultural diversity is found. These issues of disagreement include, but are not limited to, ways of protection, whether through the development of databases or through some regulatory instrument(s) and the status of any such regulatory instrument(s) (legally or non-legally binding, if parties agree for the development of an instrument or instruments); and the requirements of disclosures in patent applications (disclosure of PIC, source or origin of the genetic resources and disclosure of evidence of benefit-sharing arrangements), where the development of the product or process for which the patent is sought involved the genetic resources and perhaps the TK of local and indigenous communities. Until now, delegates to the meetings of the IGC have not really agreed on these issues, despite achieving what was considered to be the breakthrough at the Eighth Meeting of the IGC when delegates agreed on a parallel track procedure for the protection of TK[52] and cultural expressions[53] based on two sets of draft provisions. Moreover, at IGC 10, parties considered some practical options that would provide an international dimension to the draft provisions.[54]

From our perspective, the issues over which the developed and developing countries disagree in the IGC processes are linked to the rights-related concerns of local and indigenous communities identified earlier. Of course, showing evidence of negotiations of benefit-sharing contracts with local communities as a prerequisite for filing a patent application and actually providing such communities with their share of the benefits deriving from the commercial exploitation of their assets are two different things. Nonetheless, in order to provide the communities with the incentive to pursue conservation and sustainable use of biological and genetic resources at the local level, and to continue to preserve and share their knowledge freely, evidence of respect to their right to benefit in a patent application can contribute significantly. The same analysis applies in respect of the communities' rights of free and prior informed consent which should also be added as a requirement for patentability to be crafted in some way as a requirement for a patent application. If the IGC fails to include these community rights concerns in the provisions related to the protection of TK under development, then community rights are seriously undermined within WIPO processes.

One positive element is, at least theoretically, the approach adopted at the WIPO to ensure participation of indigenous and local communities to the meetings of the IGC. Indigenous and local communities' participation was enhanced by the WIPO in order to give them the opportunity to have their say in the proceedings of the IGC as they highly impact on their lives. Over 130 non-governmental organizations (NGOs) representing indigenous and local communities have been accredited by the WIPO and can attend the meetings of the IGC. In 2005, a voluntary fund was established by WIPO, where members and other interested donors can put money which will be used to support the participation of representatives of accredited groups of indigenous and local communities. However, it is questionable whether the enhanced participation of indigenous and local communities can be equated to the quality and effectiveness of their participation. In other words, it is not clear how indigenous and local communities can influence the debates in the IGC meetings, with no mastering of IP subjects and less familiarity with international negotiations. Indeed, as described by Johanna Gibson, representatives of accredited indigenous and local community groups are increasingly frustrated by their ineffective participation at the IGC meetings, as they complained over their marginalization from the main discussions during the thirteenth session of the IGC held in Geneva from 13 to 17 October 2008. In a fashion similar to the manner in which indigenous people's representatives complained about their participation in the processes of the Ad-Hoc Open Ended Working Group on Protected Areas in the context of the CBD processes, indigenous peoples' participation in the WIPO process is ineffective. These peoples are frustrated that they cannot express their views at critical moments of the debates, making it highly unlikely for the conclusions of the WIPO IGC meetings to be based on or to reflect their expectations.[55]

Farmers' rights in the FAO international undertaking and the IT-PGRFA

The concept of farmers' rights was first mooted, albeit vaguely, in the international undertaking (IU) on Plant Genetic Resources, which was adopted at the twenty-second session of the FAO conference in November 1983. As adopted in 1983, the IU instituted the principle of universal availability of all genetic resources to be accessed and used by mankind without restriction. At its adoption, the IU did not feature the recognition and protection of PBRs, which was the basis of reservation by some countries (largely developed countries) to support its adoption. This reservation was based on developed countries' perceptions that the IU was running counter to their obligations to provide legal recognition and protection of PBRs at the national level, arising from their engagement in the UPOV conventions (then UPOV 1961 and 1978). But it is not only PBRs that did not feature in the original version of the IU. Even farmers' rights were not included in the 1983 IU to the discontent of developing countries and farmers' associations, who were concerned that the universality principle will very simply leave them at the mercy of corporate users of traditional germplasm, who are unconcerned about providing any compensation to source countries. However, at the FAO conference in 1989, and in order to appease all parties, the FAO General Assembly arrived at a number of compromises. First, to appease the countries concerned with the lack of recognition of PBRs in the IU, Resolution 4/89 was adopted. This Resolution accepted the primacy of PBRs and allowed member countries bound by UPOV to impose very limited restrictions on farmers' practices of the free exchange of seeds and other cultivating materials as provided by the IU. Second, the FAO conference had to appease member countries that accepted the inclusion of PBRs in the IU by making some provisions on farmers' rights. This was done rather vaguely through Resolution 5/89 which recognized the contribution and rights of farmers to the maintenance of agro-biodiversity.

Despite making it incumbent to countries to legislate on farmers' rights in their domestic regulatory instruments, the 2001 ITPGR for Food and Agriculture is the international instrument that provides a strong and elaborate shape to the concept of farmers' rights.[56] The ITPGR deals with farmers' rights in Article 9. The Treaty recognizes the contribution of farmers to the conservation and development of PGRs and entrusts countries with the responsibility to realize farmers' rights at the national level. Critical aspects of farmers' rights needing realization at the national level include:

- the protection and promotion of TK relevant to PGRFA;
- the protection and promotion of the rights of farmers to participate equitably in the sharing of benefits arising from the utilization of PGRFA; and
- the rights to participate in making decisions at the national level with respect to the conservation and sustainable use of PGRFA.

Farmers' rights are part of the broader concept of community rights as defined earlier in the first chapter and discussed throughout this book. When we think of the way in which farming communities pursue their livelihoods, their farming and agricultural practices are the strong foundation upon which their livelihoods are rooted. Access to, exchange, and handling of planting materials are critical for the smooth pursuit of farming communities' agricultural practices from which they derive food for their households' consumption. Over the millennia, through their agricultural practices, farmers have contributed to the improvement of agricultural biodiversity, which is crucial for food security at local level and beyond.

The concept of farmers' rights is about recognizing farmers' contribution to the maintenance of agricultural biodiversity through the pursuit of their self-developed agricultural innovations and sustainable practices. According to Regine Andersen, farmers' rights consist of the customary rights that farmers hold as stewards and innovators of agro-biodiversity since the early days of agriculture to save, grow, exchange and maintain plant varieties. Furthermore, these rights consist of the legitimate rights of farmers to be rewarded and supported for their contribution to the global pool of genetic resources as well as to the development of commercial varieties of plants, and to participate in decision making on issues that are likely to affect their lives.[57] This approach to defining farmers' rights emphasizes the fact that instead of being understood or considered as the owners of traditional landraces employed in their farming practices, the concept works best in the interest of farmers when they are considered rather as stewards of their innovations and practices associated with their traditional landraces. The basis for the stewardship option lies with the fact that the ownership approach is simply contrary to communities' traditional ways of life, which promote the common ownership of their TK associated with their seeds and other agricultural materials and the free exchange of these materials among themselves. The stewardship approach to the promotion and protection of farmers' rights will avoid complications over issues as the identification of who is entitled to benefit – on an individual basis – at the community level, if not the community as a whole. Also, it can be argued that if the regulatory approaches applied in the promotion and protection of farmers' rights were based on the recognition of farmers' individual ownership of their traditional germplasm then, in the hope and expectation of corporate interest, this could disincentivize farmers to freely share the resources, and any benefits arising from corporate use would accrue to the owners only. owners only. Measures to promote and protect farmers' rights based on the ownership approach could therefore be similar or come close to private IP approaches such as patents or the UPOV PVP systems. The risk for including these private IP approaches into old-fashioned community systems of managing assets is that they could undermine community contribution to agro-biodiversity by minimizing the free share and exchange of the materials and thus could have an impact on food security at the community level.

Conclusion

This chapter demonstrates that community rights do not enjoy the same international strengths and backing as the private monopoly rights such as patents and PBRs systems. Patents and PBRs are strongly regulated by legally binding treaties such as the TRIPS agreement of the WTO and must be complied with at one point or another and in the same manner by all members of the WTO. In addition, PBRs are internationally regulated through the UPOV system, which is portrayed by proponents of PBRs (seed industries mostly based in the developed countries) as the *sui generis* system for the protection of plant varieties contemplated in the TRIPS agreement. As community rights do not earn the same strength of international protection and industry backing it is thus incumbent upon developing and biodiversity-rich countries to start legislating at the domestic level in a way that crystallizes their seriousness to take in their own hands the concerns of indigenous and local communities about access to and scientific and commercial exploitation of their assets. One way of doing so is through the establishment of national regulatory measures which address indigenous and local communities' concerns, from a human-rights perspective.

3 Brief overview on community rights in selected national regulatory instruments

As identified in the previous chapters, in some developing countries and countries rich in biological and cultural diversity there are national regulations with provisions on the rights of indigenous and local communities. In broader terms, these regulations are established as an expression of these countries' commitment to fulfil their obligations under international law. In addition, countries establishing national regulations addressing community rights do so because they want to address the concerns of their communities and to protect national interests in respect of conservation and sustainable use of their resources. Generally speaking, issues such as the various benefits that the country might glean through regulated access to its biological resources and the associated knowledge held by its traditional and indigenous communities are at the core.

This chapter examines the foundation of national biodiversity regulations within international law, stressing in particular the 1992 CBD as the main international treaty which sets the basic international obligations for all parties in respect of biodiversity conservation, sustainable use of biological resources, benefit sharing and community rights. Furthermore, it discusses national approaches adopted by some of our study countries and others in addressing community rights at the domestic level.

Foundation and features of national biodiversity regulations

The 1992 CBD[1] is a landmark international treaty that addresses the issue of global biodiversity conservation.[2] The history of the development and features of the ABS regulations, increasingly being established in several biodiversity-rich countries, can be traced back from the process of negotiation of an international treaty on global biodiversity conservation started in 1987 and culminating with the development of the 1992 CBD in Rio de Janeiro, Brazil.[3] As discussed by Graham Dutfield, perhaps none of the pioneering actors to the development of the CBD, the USA and the IUCN (International Union Conservation of Nature or the World Conservation Union) had anticipated that their efforts would yield an international treaty whose scope would go beyond biodiversity conservation to include biotechnology, technology transfer and IPRs, especially patents.[4] The Philippines was one of the first countries to enact a national ABS regulatory instrument, which, as far back as in 1995, already incorporated IP-related provisions, especially patent-related provisions.[5]

The CBD covers three objectives:[6] biodiversity conservation; the sustainable use of biodiversity components; and the sharing of benefits arising out of the utilization of biodiversity components. To assist parties in the realization of these objectives, the CBD recognizes the sovereign rights[7] of states to determine through national legislation the conditions for access to the biological resourses and in their territories. In this respect, the CBD introduced some concepts that can help to streamline the relationships between countries of origin of PGRs and countries seeking access to these elements. Some of these concepts are the PIC[8] of the party providing access and 'mutually agreed terms' between the party providing access and the party seeking access.[9] Interestingly, the CBD recognizes the relevance of the TK, innovations and practices of rural communities for the conservation of biodiversity and encourages parties to respect, preserve, maintain and ensure wide application of such knowledge and practices with the approval and involvement of traditional communities who should also receive equitable shares of the benefits derived from the utilization of their knowledge, innovations and practices.[10] Not least, the CBD recognizes the importance of IPRs, stressing in particular patents within the broader framework of access to, use and exploitation of biological and genetic resources, including the development of relevant technological processes and the possibility of transferring the relevant technology to developing countries and countries rich in biological and cultural diversity. The Convention recognizes that patents are likely to impact on the implementation of its objectives, thus encouraging parties to cooperate with the aim of ensuring that patents and other IPRs are not counter-productive but are rather supportive to its objectives.[11] The inclusion of biotechnology and IPRs (patents) in the text of the CBD therefore represents the justifying foundation of developing countries' heightened consideration and incorporation of patents and other IPRs in national biodiversity legislations which are purposely established to operationalize the CBD at the domestic level.

The CBD provides the international legal foundation and framework for the development of biodiversity legislation at the national level, to regulate access to and use of PGRs and to address the protection of the rights of rural communities.[12] Some countries have designed their biodiversity regulations to address access to and subsequent use of PGRs, TK and the protection of the related customary rights of rural communities within the framework of one of the standard forms of IP systems such as the patent system. For example, the Biodiversity Law of Costa Rica stresses that the protection of genetic resources and associated TK should be contemplated within the framework of the standard IP laws.[13] The offices administering IP laws, notably the National Seed Office and the Intellectual Property Office, must consult with the technical office of the National Commission for the Management of Biodiversity (CONAGEBIO) prior to granting any patents over any inventions based on genetic resources and TK. Furthermore and in respect of what needs to be regulated and protected by the national legislation, some countries have gone on to specifically target the assets of rural and local communities. Thus, in the cases of Panama,[14] Peru,[15] Brazil,[16] and India[17] and of South Africa, Kenya and Ethiopia in the African context,[18] national biodiversity regulations have identified the bio-cultural assets of local and rural communities as 'protectable' subject matters. Theoretically, this approach seems to address the concerns and expectations of local communities in a more direct manner which would seem to be satisfactory to these peoples. But it is a different question as to whether this approach works in practice in terms of tackling biopiracy and enhancing local peoples' trusts in bio-prospecting activities on the one hand and dealing with sustainability issues on the other.

As it transpires from the above, key features of national biodiversity legislations include, but are not limited to, their provisions on the subject matters that are regulated and subject for protection. Other important features of national biodiversity legislations are their provisions on the human-rights-related concerns of indigenous and local communities such as their rights to free and prior informed consent, their right to benefit from the utilization of their assets and their right of participation in the decision-making processes pertaining to access to and use of their assets. In addition and increasingly, the right of local communities to pursue legal proceedings against any entity that trespasses their rights is crafted under biodiversity-related laws, albeit in a vague and unpractical manner (as we see below). Some (not all) of these features are examined in turn based on some of the above-mentioned regulations.

Protectable subject matters, IP issues and benefit sharing

Countries adopt varying approaches in their designation of the protectable subject matters in their respective biodiversity-related laws. Articles 3–6 of the Law of Panama address the protectable subject matters and the related 'registrable' rights and consider the TK of local and indigenous communities as the only protectable subject matters. Rural communities can register their collective rights over TK in one of the offices mentioned in Article 4, namely the General Office for the Registry of Intellectual Property of the Ministry of Commerce and Industry or the National Copyright Office of the Ministry of Education. However, the Brazilian, Peruvian (which is a TK-specific law) and Indian biodiversity laws consider both the genetic resources and TK as protectable subject matters.[19] Some of the above-mentioned biodiversity legislations, notably the Brazilian Provisional Measure, the Peruvian and Panama laws have provisions on enforcement, especially addressing acts of infringement over the communities' collective rights.[20] For example, Article 30 of the Brazilian Provisional Measure provides that any acts of omission that contravenes the terms of the Provisional Measure should be considered administrative offences and should be punished with sanctions as varied as, among others, a warning, a fine or confiscation of a sample of components of the genetic heritage or of the instruments used for collecting and processing them. The Provisional Measure entrusts the Management Council-created pursuant to Article 10 and empowered with many responsibilities listed in Article 11 – with the authority to determine the sanctions and decide upon the fate of the samples at stake. Therefore, from the Brazilian perspective, it does not seem to be the case that disrespect of community rights in the course of access, such as failure by potential users of TK to seek PIC, is an offence that must be punished. Also, the Brazilian approach does not appear to give local and indigenous communities any role in the process of identification of acts of infringement and subsequently in challenging the offenders through the court for instance. The similar trend of preventing local and indigenous communities to challenge users of their assets in disrespect of their rights is replicated in the Peruvian law.

Two of the objectives of the Peruvian law are to ensure that the use of the knowledge of local and indigenous communities takes place with the PIC of the indigenous peoples and to avoid situations where patents are granted for inventions made or developed on the basis of collective knowledge of the indigenous peoples of Peru without any account being taken of that knowledge as prior art in the examination of the novelty and inventiveness of the said invention.[21] Despite these strongly worded provisions, which imply some degree of seriousness in the legal recognition of the human rights to PIC of indigenous communities by the Peruvian regime for the protection of TK, it remains the case that the regime does not empower these communities to directly challenge those who trespass upon their rights through the normal judicial system. Articles 47–62 of the Peruvian law, which address the handling of acts of infringement, lay down an administrative system in which the role of indigenous communities is typically limited to notifying the Office of Inventions and New Techonologies (INDECOPI) through their representative organizations of a possible act of infringement. The remaining aspects of the judicial processes are largely handled by INDECOPI with very questionable participation of the communities concerned.

The same trend of vagueness or rather failure of existing laws to empower local and indigenous communities to launch judicial challenges in terms of enforcing their rights through the courts is observed in the cases of India, South Africa and Brazil. Generally speaking, none of these countries provides in its biodiversity regulations that local and indigenous communities have the rights to seek legal remedies through the courts or through other legally recognized means of dispute settlement procedures and institutions against any actor who in the course of access to and use of their assets infringes upon their rights. Rather, in general terms, current laws in Brazil, India and South Africa have essentially defined offences and penalties in respect of acts that largely contravene the legal procedure for access (the permitting system) and use of the biological resources and associated TK. For instance, in the case of Brazil, Chapter VIII of the Provisional Measure deals with administrative offences and penalties. Article 30 of the Provisional Measure defines an administrative offence against the genetic resources and associated TK as any act or omission that contravenes the terms of the Provisional Measure and other relevant legal provisions. Further down, Article 30 provides a list of the possible punishments associated with acts amounting to administrative offences. Such punishments include, but are not limited to, a warning, fine, confiscation of samples of components of the genetic heritage and of the instruments used for collecting or processing them, or of the products obtained on the basis of information relating to associated TK, a ban on activities, suspension of registration, patent, licence, authorization; cancellation of registration, patent, licence, authorization; and loss or suspension of financing arrangements with an official credit establishment. From a human-rights angle, one striking lack from this chapter on administrative offences and penalties is a key requirement that local and indigenous communities can rightfully seek legal remedies through the courts against any actor who trespasses upon their rights in the course of access to and subsequent use of their assets. In Chapters 7 and 8, we suggest and discuss avenues which can be pursued in order to empower local communities to initiate and pursue judicial challenges. These avenues are grounded on human-rights principles.

Some of the existing initiatives have also dealt with their relationships with the standard IP regimes. Article 31 of the Brazilian Provisional Measure provides that any application for patent protection over an invention based on genetic resources and/or TK should disclose the origin of the knowledge or the genetic material. However, Article 8(4) of the Brazilian Provisional Measure clarifies that the protection provided by the law should not prejudice or limit the standard IPRs over these materials. This seems to indicate that if patent rights have been granted to an inventor who might have used TK and/or biogenetic materials, patent rights have an edge over communities and state-recognized rights to the raw genetic resources and TK under the Provisional Measure and these registered rights may not even affect the validity of such a patent. However, the Brazilian Provisional Measure has an interesting approach in the regulation of biodiversity research partnership, which can prove useful in the improvement of the technological capacities of the country via technology transfer. Indeed, Article 16(7) of the Brazilian Provisional Measure provides that research on the components of the genetic heritage of the country shall be 'preferably' done in the national territory. Arguably, in this manner countries with less technological capabilities but who provide the resources to technologically capacitated countries will earn greater benefits through better access to technology.

Benefit-sharing issues are among the key elements that are regulated by and represent critical features of current biodiversity legislations. In the case of Brazil, the General Provisions of the Provisional Measure provide for the rights and obligations concerning among other issues, the fair and equitable sharing of the benefits derived from the exploitation of the components of the genetic heritage and the associated TK. The general provisions therefore signal benefit sharing as the key component of the Provisional Measure, which it treats in more details in its Chapter VII. In dealing with benefit sharing, the Provisional Measure, however, assumes that access to the genetic heritage and TK will be sought by overseas users and exploitation/transformation of these assets will happen outside of Brazil. The Provisional Measure provides that exploitation of the genetic resources and TK shall be agreed on a contract for the use of the genetic resources and benefit sharing. This contract must indeed expand on the agreed benefits and must clearly identify the contracting parties who may be the representative of the indigenous community and the official indigenous body or the representative of the local community or the national institution authorized to have access and the receiving institution. However, in terms of addressing the rights-related concerns of local communities in respect of benefit sharing, it is our view that contracts must provide that benefits deriving from such exploitation should be shared equitably between the contracting parties, among whom, as stated above, local and indigenous communities should be the rightful recipients of the benefits because they are the contracting parties in the contract for the use of genetic heritage and benefit sharing.[22]

Understandably, in providing for the legal protection of the rights of rural communities over their genetic resources, craft products and/or the associated TK as registrable rights, these national instruments have advanced the debates about the legal protection of community rights. However, the major weaknesses of these laws include the limited participation of rural communities in the decision-making processes pertaining to access to biological and genetic resources and the absence of practical guidance for rural communities to enforce their rights before the relevant judicial authorities. After this brief discussion on the broader features of biodiversity regulations as they are currently shaped in some national jurisdictions, we now turn to exploring other aspects of these instruments in our study countries.

PIC of local and indigenous communities in the Peruvian ABS legislation

Central to the expectations of indigenous and local communities over the regulation of access to, acquisition of, and exploitation of their assets by outsiders is the fact that biodiversity regulations need to accommodate the requirements for their free and prior informed consent in the relevant processes. In other words, those seeking access to the assets of local communities must seek permission from the said communities in an open and truly honest manner. In effect, indigenous and local communities expect the law to require that all stakeholders involved in access processes in whatever capacity should make sure that prior to the actual act of access to the community assets, the latter are comprehensively informed about the purpose of such access, the intended use of the resources, and the benefit that they should expect after access has been granted and exploitation has occurred according to the relevant access authorization permits. The development and possible implementation of improved and workable regulatory measures in Peru for the protection of community rights over PGRs and associated TK, innovations and practices must not depart from these aims if these measures are intended to address the concerns of local communities from a human-rights angle.

There are indeed strong legal precedents favouring the inclusion of the PIC of indigenous and local communities in a domestic law regulating access to and use of TK in Peru. The first one is the fact that Peru is a member of the Andean Community, which is bound by Decision 391 adopted in 1996 by member countries of this community, as a common regime governing access to genetic resources. Decision 391 is binding to all member counties of the Andean Community. This decision goes beyond the ambit of Article 8(j) of the CBD, to recognize and subject to national law the rights of indigenous peoples to control the use of their knowledge, innovations and practices.[23] Understandably therefore, it follows that, with their rights to control access to and exploitation of their knowledge, innovations and practices, indigenous and local communities must be rightfully informed before access and exploitation by outsiders happen. Alongside Decision 391 of the Andean Community, but this time at the domestic level, another important legal precedent is the adoption in July 1997 of Ley No. 26839, which is the Legislation on the Conservation and Sustainable Use of Biological Resources. Following the trend and the spirit of Decision 391 of the Andean Community, Ley No. 26839 also provides that the knowledge, innovations and practices of indigenous communities over biological diversity are their cultural patrimony over which they hold rights and can decide upon use. Holding the rights to decide upon the use of their biological resources and associated TK, innovations and practices thus entitles the communities their rights of PIC in respect of access to and use of their assets by outsiders, mostly academic and corporate and business-oriented users.

Another very critical piece of legislation in Peru addressing the concept of PIC is Ley No. 27811 of 2002: Introducing a Protection Regime for Collective Knowledge of Indigenous Peoples Derived from Biological Resources. This law formalizes the requirement for PIC of indigenous and local communities in respect of access to their assets. This formalization can be evidenced from various angles. The law starts off by providing an expanded definition of the concept of PIC. It contends that PIC means:

> Authorization given under its protection regime by the representative organization of the indigenous peoples possessing collective knowledge and in accordance with the provisions recognized by them, for the conduct of a particular activity that entails access to and use of the said collective knowledge, subject to the provision of sufficient information on the purposes, risks or implications of the said activity, including any uses that might be made of the knowledge, and where applicable on its value.[24]

This definition does well to stress that the peoples whose informed consent is needed are the indigenous communities, and their PIC can be sought and obtained through their representative organizations. Furthermore, this definition enumerates areas where information is needed, notably information on the purposes, risks or implications of the activity. It can therefore be assumed that to be consistent with the law and to address the concerns of indigenous communities, the PIC process that must be complied with in order to operationalize objective (d) of the law[25] and as required under Article 6 (conditions of access to collective knowledge), needs to provide sufficient information to the satisfaction of indigenous and local communities. In order to step up the fight against bio-piracy in Peru, the regional government of Cusco (the Peruvian Andes) introduced a law in January 2009 specifically aimed at empowering local communities to control access to their assets at their level and to be compensated properly when they approved access to their assets according to the law. As reported by www.scidev.net/en, the Cusco government is claiming to be the first to establish a regional and locally applicable law in the world to outlaw bio-piracy.[26] The Cusco law empowers local communities to implement ways to protect their assets, including through the development of community registers and community protocols to guide their interactions with scientific and corporate actors. With this law, scientists and corporations must seek permission and deal with local communities directly and, subsequently, must share benefits with them. This law has not been implemented, so it is hard to judge its effectiveness in safeguarding that community PIC is really sought during access processes. Nonetheless, it appears to be a step forward in empowering, from a practical angle, local communities to exercise control over access to their assets at the local level.

The analysis of the 'bio-piracy' case studies in Chapter 5 will test the breath of the PIC process in respect of the information provided by seekers of access during the PIC and consultation process and the magnitude of satisfaction of rural communities about this information. Ascertaining that the cases analysed amount to 'bio-piracy' will also depend on the effectiveness of the PIC process.

Access procedures and permitting approach: community rights of participation in South Africa

Broadly speaking, access to indigenous resources and associated TK in South Africa is regulated under the National Environmental Management Biodiversity Act No. 10 of 2004[27] and the related implementing regulations on Bioprospecting and ABS,[28] which were adopted in February 2008 and which entered into force in April 2008. While the Biodiversity Act No. 10 was very broad in its scope covering a diverse range of issues such as the creation of the South African National Biodiversity Institute (SANBI), biodiversity planning and monitoring, and threatened or protected ecosystems and species, Chapters 6 and Chapter 7 of the Act specifically dealt with Bioprospecting and ABS and the procedures for access through the permit system. The provisions of Act related to bio-prospecting, access and export of indigenous resources and the procedures leading to the acquisition of an access authorization permit were very broad and had very limited practical significance in respect of their implementation. For instance, based on the provisions of Chapters 6 and 7 of the Biodiversity Act No. 10, it was not clear what kind of information actors (e.g., bio-prospectors and academic researchers) seeking access to indigenous resources were expected to provide to relevant stakeholders in respect of the PIC process. Furthermore, the Act was surprisingly cloudy as to when precisely the local and indigenous communities will have to intervene in the access process and why their involvement in the process is critical, especially when the overall aim is for other actors to eventually acquire community assets. With these limitations, some sections of the Act such as Chapter 6 and the related issues were put on hold in respect of their entry into force until specific regulations with more practical effects were developed and enacted. The Commencement of the Regulations on Bioprospecting and ABS in April 2008 marked the effective entry into force of Chapter 6 of the Biodiversity Act No. 10

of 2004 and its related provisions. As briefly explained below, there is no doubt that the Act and its implementing 2008 regulations on Bio-prospecting and ABS have done well in streamlining and harmonizing the access procedures as compared to when access to community resources were regulated in a very fragmented fashion under the old regulatory framework. However, despite such improvement in the regulatory landscape, the 2008 Regulations have not really ensured substantial participation of local and indigenous communities in the decision-making process pertaining to access to and use of their assets. At the very best, the new regulatory landscape can be praised for enshrining such requirements as the PIC and the need for benefit-sharing evidence in the process of access, which are all issues that are currently debated in the context of the development of the international regime and in the context of the meetings of the WTO council of TRIPS on genetic resources and the protection of TK. In a sense, one can arguably state that in crafting these issues in national regulations, the South African legislator is indeed implementing the country's right of national sovereignty to regulate access to the resources found in its geographical territory on the one hand, while fulfilling its obligations under international law on the other. But the rights of local and indigenous communities to fully take part in the decision-making processes on issues of interests to them enshrined in the 2007 UN-DRIP are still heavily undermined in the relevant access processes. This failure of Act No. 10 of 2004 and the 2008 implementing regulations on ABS is a continuation of failure by the previous regulatory landscape and permitting process to include local and indigenous communities in the decision-making process.

Prior to the enactment and entry into force of the Biodiversity Act No. 10 of 2004 and its 2008 Regulations on Bio-prospecting and ABS, access to biological and indigenous resources for scientific or commercial purposes was still subject to the acquisition of a permit or licence issued by provincial conservation departments under the old regulatory landscape. Each provincial conservation department had its own set of procedures, although there were similarities in some respects. Three instances for access to, collection of and acquisition of indigenous resources for research or other purposes required the application of a permit at the provincial level:[29]

1. Where the applicant states that they want to collect biological materials for research purposes.
2. Where the applicant states that the collection is purely for personal, non-commercial and non-research purposes.
3. Where the applicant states that the intention is to collect for commercial purposes.

With regards to plant collection for research purposes, whether or not the applicant was a South African citizen, in order to be granted a research or plant collection permit, they should have been affiliated with a South African institution. The application pack was expected to include the research proposal detailing the species that were to be collected, the quantity and the area where collection would take place. During field collection, the researcher was expected to be accompanied by a senior staff member from the provincial nature conservation office. After collection had been carried out, the permit applicant was expected to provide the department with duplicates of voucher specimens and copies of their research results.[30] Depending on each provincial nature conservation department, the permit was issued either by the head of the departmental nature conservation office or by the head of the regional and district nature conservation offices.[31]

Following the format described above, the South African permit system was patchy, characterized by the lack of harmony among all the provincial nature conservation departments. This lack of harmonization was not helpful for the purposes of biodiversity conservation because different access procedures cannot ensure effective monitoring of access to and collection of PGRs throughout the national territory. Furthermore, the mechanisms employed in the examination of applications for research or plant collection permits and ultimately the grant of such legal authorizations in the various provinces did not provide for the full participation of the concerned rural communities. Indeed, as long as the applicants fulfilled the conditions set out by the departmental nature conservation offices, they legally expected to be granted the research or plant collection permit without having to go through any consultative process in the course of which the consent of rural communities would have been sought and acquired.

However, as hinted earlier, if the legislator in South Africa has succeeded in achieving an outcome in the new regulatory landscape on bio-prospecting and access to biological and indigenous resources and benefit-sharing issues, it is that the current regulatory landscape has absorbed the patchiness of the previous regime. For one thing, rather than leaving key responsibilities in respect of the examination, and grant of permits in the hands of the respective provincial conservation boards, only the minister is responsible for issuing bio-prospecting permits for research to be conducted on South African indigenous resources. The minister is also the sole authority to issue integrated bio-prospecting permits for research and export of South African indigenous resources.[32] Furthermore, for research activities other than bio-prospecting to be undertaken on indigenous resources, the Member of Executive Council (MEC) for environmental affairs has been identified as the issuing authority for research and integrated research permits.[33] Interestingly, there are provisions for the delegation of power by the minister in respect of bio-prospecting activities and by the MEC in respect of non-bio-prospecting activities.[34] Despite the possibility of power delegation, the patchiness of the access regulatory process as it existed across the provincial conservation boards has been significantly reduced.

While this marks good progress, the Biodiversity Act No. 10 of 2004 and its 2008 implementing regulations have not provided that local and indigenous communities are the rightful participants in the decision-making processes pertaining to the issuance of the relevant permits that essentially allow access to their assets. For example, Section 82 of the Biodiversity Act No. 10 of 2004 and Regulations 8(1)(a)–8(1)(d) of the 2008 implementing regulations have provided for the acquisition of the PIC of local and indigenous communities as an additional requirement in the permit system. In effect, it is contended – at least theoretically – that the minister or MEC will not issue the permits if applicants have not provided evidence that they sought and acquired the PIC of the communities concerned. But our criticism of the way PIC is enshrined in the Act and the 2008 implementing regulations is that these instruments seem to in fact have secured a de-facto PIC of rural communities and private landowners to permit seekers, because the Act does not give the communities and private landowners the opportunity to accept or refuse the activities before the final approval of the access permit is issued by the relevant authority. There is no provision that formalizes local and indigenous communities' rights to say 'No' to these activities because the provisions are shaped in such a way that these people are ultimately expected to provide their PIC and to sign the benefit-sharing agreements. In addition, the Act and the 2008 implementing regulations are very silent on the need for carrying out consultations between the issuing authority and the communities concerned as a way to cross-check the validity of the proof of PIC or benefit sharing that might have been provided by permit applicants. Such consultations between the issuing authorities and the communities would really have been the platform through which the authorities might have ascertained that the various processes had been thorough and that permits can be granted because communities had rightfully provided their informed consent.

The right of participation of local and indigenous communities is therefore very weak in the landscape of instruments regulating access to resources and issues regarding benefit sharing. Neither is this situation unique to South Africa. This situation is not unique to South Africa as it characterizes the access processes in many countries rich in biodiversity. However, the situation can improve if the legislator revisits these pieces of legislations and includes this right in them in a manner that will address the expectations of local and indigenous communities and be compliant with relevant human-rights treaties.

Farmers' rights in India

The protection of farmers' rights is enshrined in the 2004 FAO IT-PGRFA. What the Treaty does is recognize the contribution that farmers and their communities have made in the promotion of agricultural biodiversity in the past and encourage them to keep on conserving and improving PGRFA. This recognition and encouragement are materialized through the incorporation of the concept of farmers' rights in the Treaty in order to reward farmers. Farmers' rights, as framed in Article 9 of IT-PGRFA, are multifaceted. Despite its failure to define the concept of farmers' rights, the IT-PGRFA suggests avenues for their realization and entrusts parties with the responsibility to legislate on farmers' rights at the national level. As provided by the Treaty, the realization of farmers' rights must therefore cover, among other issues:

- The protection of TK relevant to PGRs for food and agriculture.
- The right to equitably participate in sharing the benefits arising from the utilization of PGRs for food and agriculture.

- The right to participate in making decisions, at the national level, on matters related to the conservation and sustainable use of PGRFA.

In addition, it is the position of the Treaty that it does not intend to limit any rights that farmers have to save, use, exchange and sell farm-saved seed/propagating material, subject to national law and as appropriate. The approach to addressing farmers' rights to reuse farm-saved seeds and propagation materials is rather complex because, in effect, the Treaty does not really affirm these rights. However, the complexity also provides an opportunity for varying interpretations of Article 9(3) meaning that it is entirely up to countries to frame their domestic regulations on farmers' rights as they see fit and in their best interests but having due regard to their international obligations entered into in the context of other treaties. In the subsequent paragraphs we describe how, through two mechanisms, India has legislated on farmers' rights to save, use and sell seed and propagating materials and farmers' rights to benefits. One of these mechanisms is the India's Protection of Plant Varieties and Farmers' Rights Act, enacted in 2001. The second mechanism is the no-less-important Community Gene Banking and On-Farm Conservation in India. These two mechanisms have been dubbed 'success stories' in the realization of farmers' rights in the Indian context.[35] As success stories, Regine Anderson of the Norwegian Think Tank, the Fridtjof Nansen Institute (FNI), refers to any activity, initiatives or projects that have resulted in concrete achievements in addressing the suggestions related to the realization of the concept of farmers' rights stipulated under Article 9 of the FAO International Treaty.[36] These success stories have been gathered in the context of the Farmers' Rights Project implemented by the Fridtjof Nansen Institute as a contribution to the call issued by the governing body of the international treaty at its second session, which invited members of the FAO international treaty and relevant bodies to experiences on the realization of farmers' rights and submit their views to the secretariat of the governing body, which will use them as an agenda item for the third session of the governing body of the international treaty scheduled for June 2009 in Tunis, Tunisia.[37]

In 2001, the Indian parliament enacted a law addressing the protection of PBRs and farmers' rights entitled The Protection of Plant Varieties and Farmers' Rights Act 2001. This particular Act pursues four objectives, notably:

1. The establishment of an effective system for the protection of plant varieties and the rights of farmers and plant breeders, to encourage the development of new plant varieties.
2. The recognition and protection of the rights of farmers in respect of their contribution made at any time in conserving, improving and making available PGRs for the development of new plant varieties.
3. The protection of breeders' rights, to stimulate investment for research and development, in both the public and private sectors, for the development of new plant varieties.

4. The facilitation of growth in the seed industry in the country, which will ensure the availability of high quality seeds and plant material to the farmers.

An appraisal of these objectives reveals that they are very broad as they cover wide-ranging issues such as the seed industry and research attempts to accommodate the interests of various groups including (1) the farming communities that the Act recognizes as breeders in their own capacity and (2) technologically capacitated seed breeders. The Act covers nine specific rights of farmers among which are farmers' rights to seed, their right to reward and recognition and their right to benefit sharing.[38] Of particular relevance to this book are farmers' rights to seed, which are articulated to accommodate their rights to save, use, resow, exchange and sell farm produce including seed varieties protected by PBRs. However, in order to balance these rights with the legitimate interests of plant breeders, the Act stipulates that farmers are not allowed to sell breeders' protected varieties as branded packages. There are other important interests of farmers that the Act sought to protect such as the right to information, which protect farmers from crop failure if their handling of these crops were based on false or inaccurate information provided by plant breeders. As provided by the Act, should any such failure happen, farmers have the right to be compensated by the breeder who developed the new plant variety which caused the crop failure.[39]

As the Indian law demonstrates, it is possible to develop domestic regulations that balance the interests of various actors and areas while helping the country to fulfil its obligations under international law. In the Indian case, the genuine and legitimate interests of farmers and India (as a sovereign country) are balanced with the interests of breeders and the need to promote research and the establishment of a competitive seed industry. India is legally entitled to do so because as a member of the WTO, the country is bound to comply with Article 27(3)(b) of the TRIPS agreement, which requires each WTO member to enact a domestic law for the protection of plant varieties. TRIPS does not obligate WTO members to adopt the UPOV models for the protection of plant varieties, even though UPOV and other partner organizations such as the WIPO have been trying to convince WTO members (especially developing countries) that adopting UPOV is the best way to achieve compliance with the TRIPS provisions under Article 27(3)(b).

The second Indian example concerning the realization of farmers' rights relates to the understanding of the right to benefit from the use and management of their traditional landraces. In the dry land regions of India, the GREEN Foundation runs a people's programme aimed at reviving farmers' old varieties and re-enhancing the diversity in agricultural materials – especially traditional landraces – which are the pillars of traditional farming practices in India and thus the backbone of food security. Key activities being implemented through this programme are capacity-building for farmers provided by scientists so that they are able to maintain seed stores and on-farm seed banks, without external assistance. Information exchange among farmers participating in this programme is also critical to the realization of its aims. The main objective of this GREEN programme is the conservation of traditional landraces and the realization of their sustained availability, including the availability of information pertaining to their use to all the farming communities that are in need of a diversity of landraces most adapted to their farming conditions and practices. To fulfil this objective, the foundation establishes (a) seed stores as the repository of seeds provided by members and (b) field gene banks. The seed stores consist of landraces grown locally and are the back-ups to the local seed-market networks through which farmers have traditionally exchange seeds. These stores are also places where farmers meet to exchange information about the utilization of materials that they keep in the stores. They are therefore places where local materials, which are increasingly being eroded in the fields, can be found and reintroduced since they represent the materials that are best adapted to communities' needs and farming conditions. As of 2008, the GREEN foundation had established 25 community seed banks, each seed bank bringing together 15 to 20 members on average. The banks conserve about 43 varieties of finger millet, 84 varieties of paddy, 24 varieties of sorghum, 44 varieties of minor millets, 53 varieties of pulse, 14 varieties of oilseeds, 4 varieties of wheat and 116 varieties of vegetables. As indicated earlier, capacity-building is an important component of this programme. To address this element, GREEN pursues a participatory approach in the establishment of seed banks. With the assistance of scientists, farmers determine the criteria for selection of varieties and the varieties that meet the identified criteria for their storage in the seed banks. Most common criteria guiding the selection process include level of resistance to pest and diseases and tolerance of drought. Varietal selection is immediately followed and complemented by on-farm trials that must be carried out prior to wider dissemination of the varieties. It must be stressed that other farmers who are not taking part in this programme are having access, albeit informally, to the varieties conserved through the programme.

The benefit sharing arrived at in this example are not necessarily financial. They are benefits in terms of community access to more diverse varieties and crops germplasm adapted to local conditions; access to information concerning the handling of these germplasms; and capacity-building that can ensure continuity and sustainability of this kind of initiative. All these benefits are critically important and will undoubtedly contribute to conservation of landraces and to food security at the local level.

Conclusion

It is fair to say that national processes and mechanisms addressing issues of interest to local and indigenous communities are still very slow in properly legislating on community rights in a manner that, at least theoretically, meets community expectations. Among the examples discussed in this chapter, one can arguably say that India has addressed farmers' rights at a domestic level with strongly worded provisions but also balanced these provisions with the interests of seed breeders. However, even in the case of India, their legislation on farmers' rights does recognize that the motivation for legislating on them in this manner should be based on the internationally recognized human rights of farmers as crafted in international treaties and declarations such as the 2007 United Nation Declaration on the Rights of Indigenous Communities. Ideally, national regulatory instruments should accommodate indigenous peoples' and farmers' rights from a human-rights perspective, as this is the approach that will enable them to keep their commitment to sustainability and the free sharing and exchange of their resources and knowledge which is essential to the promotion of agro-biodiversity and food security at the local level.

4 Selected international and regional human-rights instruments and their provisions on community rights and IPRs

The concept of human rights is one of those concepts in the international discourse for which there is no universally accepted definition, despite the fact that human rights are widely understood to be universal in nature. Nonetheless, one of the definitions of the human-rights concept suggests that human rights constitute any basic right or freedom to which all human beings are entitled and in whose exercise a government may not interfere (except, I would add, in exceptional circumstances). Internationally, human rights are enshrined, first, in the 1948 UDHR and its legally binding covenants, the 1966 ICCPR and the 1966 ICESCR. These two covenants transformed the advisory nature of the 1948 UDHR into two legally binding instruments with which state parties have the legal obligation to comply. These three pieces of international human-rights instruments constitute what is known as the International Bill of Rights. They encompass a cross-section of elements of human rights based on such principles as the universality, inalienability, indivisibility/interdependence and fundamentality of human rights. Having regard to these principles, it may seem plausible that there is no point in developing a specific set of rights for indigenous and local communities. Discussions in this chapter and the subsequent ones, however are based on one of the grounds for developing the 2007 UN-DRIP which is the fact that due to the specific injustices that indigenous peoples have undergone there is a need to address their human-rights-related concerns in an express manner. Also controversial are the interpretations associated with the inclusion of IPRs into human-rights instruments. Here, we refer to Article 27 of UDHR and Article 15 of the ICESCR. For some analysts, these provisions undoubtedly relate to the protection of the human rights of peoples considered as products resulting from their intellectual effort and creative minds. On the other hand, there are views that IPRs are not human rights because they are subject to recognition while human rights are fundamental rights and thus do not require any specific recognition.

This chapter does not attempt to add to the above-mentioned controversial issues. It revolves around the assumptions that community rights indeed deserve special attention in the human-rights discourse, precisely due to the specific circumstances of traditional and indigenous communities. Furthermore, it is our view in this chapter and the subsequent ones that, indeed, the provisions of Article 27 of the UDHR and Article 15 of the ICESCR refer to IPRs, thus making them integral and important components of human-rights instruments. Following these assumptions, this chapter explores the extent to which both community rights and IP provisions are crafted in a selected number of human-rights instruments. These analyses are meant to assist us in establishing that because community rights and IPRs are important components of the human-rights discourse, policy processes aimed at addressing community concerns should strongly reflect a human-rights perspective in order for them to be comprehensive, tailored and responsive to the expectations of traditional and local communities.

Community rights and IPRs in the International Bill of Rights

After the adoption of UDHR on 10 December 1948, the plan of the UN in respect of human-rights issues was to transform the advisory and rather inspirational approach of the UDHR into obligatory undertakings that would be binding to state parties. This was in fact meant to be an immediate follow-up process. In practice, however this was not the case, as the development of these binding instruments was delayed until 1953 when it was eventually launched. This process yielded two core human-rights covenants in 1966 which, together with the UDHR, constitute what is known as the International Bill of Rights. The ICCPR and the ICESCR both came into force in 1976 after the 35th nation signed these instruments.[1] The ICCPR and the ICESCR have tried to differentiate between civil and political rights on the one, such as the right to vote, freedom of speech and freedom of assembly, and the economic, social and cultural rights on the other, such as the rights to decent working conditions, fair wages and to utilize freely and fully one's natural resources. However, despite this attempt to categorize the rights in different sets, there is general consensus that human rights are indivisible, meaning that they in fact need to be taken as a whole because from a more practical angle all these categories of rights are interlinked; the exercise and enjoyment of the rights belonging to one category are highly likely to have impacts over the exercise and enjoyment of the rights categorized in the other set.

The ICCPR and the ICESCR both have provisions that are applicable to the protection of the rights of traditional and local communities which were identified earlier in Chapter 1 and which constitute the core of this work including:

- their customary ownership rights over PGRs and land;
- their rights to fully participate in the decision-making processes regarding the management of their resources;
- their rights to be consulted prior to any access and utilization of their resources;
- their rights to determine terms for access authorization;
- their rights to monitor and control access to the resources;
- their right to fairly benefit from the utilization of the resources and TK; and
- their rights to stop any activity not respecting their traditional rules and to file a lawsuit against anybody who infringes the terms under which access was allowed.

This chapter avoids an extensive and detailed analysis of all the rights covered by the provisions of these two covenants. Rather, what it tries to do is to reveal the specific provisions which, individually and together, can be seen as addressing the concerns of local communities from a human-rights perspective. Indeed, central in addressing the human-rights-related concerns of communities identified above are the following provisions of the two covenants:

- Article 1 on self-determination which is identical in both the ICCPR and the ICESCR.
- Article 2 of ICCPR on state obligations.
- Article 14 of ICCPR on equality before the court.
- Article 25 of ICCPR on participation in public affairs and non-discrimination.
- Article 25 of ICCPR on ethnic minorities.
- Article 7 of ICESCR addressing issues of working conditions, fair wages and decent living.
- Article 15 of ICESCR on IPRs and benefits from the progress of science and technology.
- Article 25 of ICESCR on the right of peoples to enjoy and utilize fully and freely their natural resources.

The above-mentioned provisions are, as hinted earlier, all individually and together very instrumental in assisting local communities in the realization of their rights vis-á-vis their systems pertaining to the management, sustainability and control over access to and commercial utilization of their assets. The two covenants provide for the rights to self-determination. By virtue of this right, all people must freely determine their political status and freely pursue their economic, social and cultural development. Additional components of this right to self-determination include the freedoms it affords to all peoples to dispose of their natural wealth and resources without prejudice of obligations arising out of international economic cooperation and its provisions on the responsibility of the State to ensure that there is full realization of this right at the national level. A practical commitment to ensure the realization of the right to self-determination in a specific context should therefore provide a human-rights platform through which all members of the society have the opportunity to participate in a meaningful manner in deciding over their governance and common future. Including the right to self-determination in legally binding multilateral human-rights treaties represented a strong indication from the international community about its seriousness regarding the realization of this right on a global plane. One argument about the *raison d'être* of this right is that it had particularly strong relevance in the colonial era, as its realization was aimed at ensuring the liberation of some States from the hinges of colonialism. Based on this reflection, the right of self-determination would not have had any more relevance with the ending of colonialism. But this is not the case because even the functioning of the Human Rights Committee of the ICCPR (HRC), the body that oversees the implementation of the ICCPR, promotes the interpretation of this right for its application beyond the narrow spectrum of colonial liberation.[2] The right to self-determination is therefore seen as one of the most instrumental type of rights that local communities have at their disposal to assert their control over access to and use of their biological resources and associated TK by outsiders for scientifically and commercially motivated purposes. In this respect, based on the self-determination approach, traditional and local communities can assert their customary ownership over the resources which are found in and accessed from their traditional lands in their bid to exercise control over the process of access and the subsequent use of their assets.

The realization of the right to self-determination by local and traditional communities will obviously be ineffective and perhaps meaningless if any approach pursued in that respect fails to address this category of right in conjunction with other human rights related issues including participation and non-discrimination and issues of compliance and enforcement. Looking at the national laws and policies regulating access to the resources that are generally found on the lands over which traditional communities claim ancestral ties, such laws and policies generally fail to provide for effective participation of local communities in decision-making processes through which authorization for access is granted. Furthermore, researchers and others seekeing access to the assets of local communities are also generally interested in acquiring the knowledge of these peoples about how they have traditionally managed and utilized their assets. Such knowledge, when acquired, can be instrumental in the technology-based transformation of these assets towards the development of scientific knowledge and the resulting products and processes over which users generally seek IP protection through patents or PBRs. Within the framework of their interaction with outsiders who are interested in both their tangible and intangible assets (their biological and genetic resources and associated TK), local and traditional societies generally give their assets away freely and easily. Their mode of giving to outsiders, which is based on their own systems of exchange of information and resources among themselves, is based on principles of reciprocity and flexibility. Taking advantage of such easy access to and acquisition of community assets, outsiders are sometimes left with the feeling of not being under any obligation to compensate the communities after exploiting their assets in profitable (financially or otherwise) undertakings. Pursuant to the implementation of the 1992 CBD, national regulations, where they exist and are in force, require that the interaction between the 'givers' (local communities for example) and the recipients (researchers and commercial users) be based on a contractual arrangement. But if and when they are established, the monitoring of compliance to these contracts is not the easiest thing that source countries (let alone their local communities) can achieve consistently. It follows that, although in pursuit of the right to self-determination local communities may well be empowered to control access to their assets, they may not be able to realize this right comprehensively if their right to be compensated from the exploitation of their assets by outsiders and their right to challenge through the court those who trespass the relevant contractual obligations are not properly constructed in a broader regulatory framework tailored to address their concerns from a human-rights perspective. The realization of all these rights can only be possible if the principle of non-discrimination filters properly into the development and implementation of the regulatory framework. Logically, it follows that, including the rights to self-determination, participation and non-discrimination of local and indigenous communities within the processes associated with their assets can be achieved through the implementation of Article 2, Article 25 and Article 26 of the ICCPR. Furthermore, Article 14 of the ICCPR stands to be very instrumental in respect of addressing the concerns of rural communities associated with the enforceability of their rights.

In respect of the possibility for rural communities to lay claims to benefit, or be compensated for the utilization of their assets on human-rights grounds, the provisions of Article 1 of both the ICCPR and the ICESCR, Article 47 of the ICCPR, and Articles 7 and 15 of the ICSECR appear to be relevant in this respect. Article 1(b) of the ICCPR and the ICESCR and Article 47 of the ICCPR make clear reference to the freedom afforded to the peoples to dispose of their wealth and natural resources (which, in this work must be understood to encompass the tangible and intangible assets of local communities). From our perspective, having the freedom to dispose of one's own resources should mean that traditional and local communities can, by rights, set the prerequisite conditions for access to and utilization of their resources by others, including, but not limited to, their expectations for proper compensation and benefit sharing. Such a rights-based freedom is also rooted upon the widely accepted idea that these peoples have thus far put hard labour into the conservation of bio-cultural diversity in their localities, which is now the subject of obvious scientific and commercially oriented interests by business-minded operators. This is where Article 7 of the ICESCR comes into play as it can be interpreted to mean that rural communities should by right benefit from their hard labour invested in the conservation and promotion of bio-cultural diversity and earn a decent standard of living from such benefits.

Another important ground for indigenous and local communities to claim benefit from the utilization of their assets within the human-rights spectrum is achievable through Article 15 of the ICESCR. This provision is very similar to Article 27 of the UDHR which are both commonly known as the IPRs provisions of the International Bill of Rights. However, it must be stressed that, while the stipulations of Article 15 are understood to represent the IP provisions of the International Bill of Rights, it emerges from the operations of the Committee on Economic, Social and Cultural Rights that the provisions of Article 15 should not be equated to human rights, precisely because IPRs are just not human rights.[3] Having provided a summary description of what is meant by human rights and the difference between human rights and IPRs in both Paragraphs 1 and 2 of General Comment No. 17 of the Committee on Economic, Social and Cultural Rights, the comment stresses that:

> Whereas the human right to benefit from the protection of the moral and material interests resulting from one's scientific, literary and artistic

> productions safeguards the personal link between authors and their creations and between peoples, communities, or other groups and their collective cultural heritage, as well as their basic material interests which are necessary to enable authors to enjoy an adequate standard of living, IP regimes primarily protect business and corporate interests and investments.

Nonetheless, it is the approach of this book, as the human right to benefit from the protection of the moral and material interests resulting from one's scientific, literary and artistic productions safeguards the link between the author and their creations, it is equally the case that IPRs can be interpreted as safeguarding the personal link between the creators and their creations, even if this link is very likely to be associated with material and financial gains as well as the promotion of business interests of corporate users.

Through the lenses of human rights, traditional and local communities are entitled to benefit from the exploitation of their assets in scientific and commercially motivated undertakings by others. Indeed, the definition of 'author' provided by the Committee on Economic, Social and Cultural Rights is broad-based, aimed at natural persons and includes both men and women. There is no exclusion of traditional and indigenous communities in the Committee's approach to define an 'author' as stipulated in General Comment No. 17. Importantly, the general comment has been even more detailed in its clarification of what should be understood as 'any scientific, literary or artistic production'. In particular, Paragraph 9 of General Comment No. 17 on this issue mentions the knowledge, innovations and practices of indigenous and local communities as the possible components of their reference to any scientific, literary or artistic production. From the perspective of this work, regardless of whether the assets of local and indigenous communities are eligible for protection under the formal patents and/or PBRs systems, they must be considered as the IPRs of these peoples, who must thus be entitled to sound benefit and compensation when they have been exploited by others.

However, modern IPRs such as patents and PBRs held by others over the assets of indigenous and local communities must also be considerate of the concerns of these peoples, such as concerns over resources depletion and impacts on the cultural integrity of the peoples. The way to guarantee that such considerations are fully integrated within the framework of IPRs is, for one instance, to make sure that they are captured by actors involved in the formulation of IPR policies. It thus appears to be the case that a human-rights approach to IP is needed. In the process of formulating IPRs policies, a human-rights approach must impose an additional standard from the classical IPR approach which favours big businesses and corporation with less regard to the interests of the wider public, let alone indigenous communities. This aspect will be addressed in Chapters 7 and 8 when we will be discussing the innovative regulatory approaches that can be taken into consideration in an attempt to establish a human-rights approach in the formulation of IPR policies, which are used by scientific and commercial actors to secure significant economic rights over the derivatives of the assets of indigenous and local communities.

Community rights in the 1989 ILO Convention No. 169

The ILO Convention No. 169 concerning indigenous and tribal peoples in independent countries was adopted on 27 June 1989 by the 76th General Conference of the ILO meeting in Geneva. This instrument is essentially the outcome of the revision of an earlier ILO Convention No. 107 on Indigenous and Tribal Populations adopted by the 40th Session of the General Conference of the ILO in 1957 in Geneva. Much has been written on the substance and timely development of ILO Convention No. 107 and on the need for its revision into ILO Convention No. 169. The purpose of this revision was to develop an instrument that would reflect developments in the international legal regime regarding issues of interests to indigenous and tribal peoples that had occurred between 1957 and 1989, including, for example, the massive decolonization of several countries hosting large populations of indigenous and tribal peoples. The history and background to the developments of these conventions is therefore beyond the scope of this work (see Thornberry 2002).[4] However, the thrust of this section is very much to address the issue of participation of indigenous and local communities in the decision-making processes on issues that are likely to affect their lives, such as access to their assets and exploitation of them in scientific and technology intensive undertakings, to the extent that such participation is structured in the body of international law. Therefore, this section focuses on ILO Convention No. 169 and its specific provisions which are relevant to indigenous people's full participation (not just involvement) in major processes pertaining to the use, management and access of natural resources for scientific and commercially motivated purposes.

Comprising 44 articles in total, the ILO Convention No. 169 covers various issues of interest to indigenous and local communities. These issues are indeed clustered in the different parts of the Convention, and they include land issues, issues related to recruitment and conditions of employment, vocational training, handicraft and rural industries, social security and health and issues related to education and means of communication. For the sake of shortness and simplicity, this brief exploration of the Convention's provisions is aimed at those provisions that are intended to promote the full and proper participation of local and rural communities in the decision-making processes on issues that affect them. From our perspective, the Convention's approach in defining the peoples at which its provisions are aimed, its stand on the values of certain valuable assets to these peoples and the critical need for all stakeholders involved in the management, policy development and use of these assets, are very relevant and applicable to the socio-economic and cultural circumstances of the peoples targeted in this work.

Article 1 of the Convention sets the scene by defining the subjects to whom its provisions apply. Article 1(1)(a) provides that the Convention applies to tribal peoples in independent countries whose social, cultural and economic conditions distinguish them from other sections of the national community, and whose status is regulated wholly or partially by their own customs or traditions or by special laws or regulations. An important criterion for the identification of indigenous or tribal peoples is provided further in Article 1(2) which stresses upon the fact that 'self-identification' as indigenous or tribal shall be regarded as a fundamental criterion for determining the groups to which the provisions of the Convention apply. Based on this Article, the peoples whom this work indifferently addresses as traditional societies, indigenous peoples, traditional or local communities fall within the purview of ILO Convention No. 169 and can defend their rights and interests using its provisions. The right of participation of local communities in the development of policies aimed at their human-rights-related concerns and the approaches the Convention suggests in respect of its realization are stipulated in a number of provisions including Article 2(1), Article 2(2)(b) and Article 4(1). These provisions cover the fundamental obligations that governments must fulfil in a bid to ensure participation of local communities in the policy processes aimed at promoting their social, economic and cultural rights based on tailored regulatory measures. The Convention does not specify these measures. However, from our perspective, such measures must include law-making processes which governments must initiate in consultation with the peoples concerned in order to ensure that the resulting instruments reflect their views and expectations. As identified in the early chapters of this work, the subject matters over which traditional and local communities have rights-related concerns fit squarely with the framework of the Convention's approach in defining the elements that are valuable to these peoples. Parts (a) and (b) of Article 5 of the Convention accommodate these matters. Following these articles, the social, cultural, religious and spiritual values of the peoples must be recognized and protected as well as their rights over the natural resources associated with their traditional lands. The Convention emphasizes that the rights contemplated in these articles must include the right of the peoples to participate in the use, management and conservation of the resources. In translating these provisions to the framework of this work, traditional germplasm (local seeds developed by traditional farming communities), PGRs and the associated TK and practices of indigenous and local communities, all of which are usually sought by scientific and corporate bodies, typically represent the valuable resources contemplated by the Convention. In this respect, the rights of local communities, including their ownership rights over these resources based on their customary laws and their right to participate in any activity such as academic and fundamental research based on these assets should be respected and protected.[5]

Issues of consultation with and participation from indigenous and local communities in the context of developing policy measures on access to and utilization of their assets in scientific and commercially motivated undertakings can be best addressed with the backing of the provisions of ILO Convention No. 169. Perhaps the very limited and rather slow ratification of this treaty by countries worldwide may be explained by the very strong tone of this convention over community-rights-related issues. However, without undermining the power of the relevant authorities in respect of the issuance of the relevant legal authorization permits to access and use the assets of local and indigenous communities, it is still possible that national systems are set up in a way that ensures substantial and effective participation of local and indigenous peoples in policy development and decision making in a manner that reflects national priorities on these issues whilst also respecting the world-views of traditional actors. The measures we suggest in Chapters 7 and 8 in respect of improving the permitting systems will reflect this approach.

The 1981 African Charter on Human Rights and the African approach to self-determination

With the signing by a group of African heads of states of the Organization of African Unity (OAN) (now the African Union [AU]) Charter in 1963, there was an urgent need to develop an African approach to the realization of human rights in order to give effect to the Universal Declaration of Human Rights in the context of Africa on the one hand and the Charter of the OAU on the other. Put in a very simple and straightforward manner, Africa embarked on a long process for the development of a human-rights instrument, which culminated with the adoption of the African Commission on Human and People's Rights (ACHPR) in 1981 in Nairobi, which eventually entered into force in 1986. Today, all African member countries of the AU have ratified the ACHPR.

As indicated above, the ACHPR is closely linked to the Charter of the OAU and the Constitutive Act of the AU, which is the body that replaced the OAU in 2000. Making a clear reference of the OAU Charter in its preamble, the ACHPR declares that the realization of the objectives of the OAU Charter in respect of freedom, equality, justice and dignity will assist in the realization of the legitimate aspirations of African peoples. It must be stressed that none of the Charter of the OAU, the Constitutive Act of the AU, or the ACHPR mention such terms as 'indigenous communities', 'traditional societies' or 'local communities'. Nonetheless, as we discuss below, there is a working group dealing with the rights of indigenous populations and communities established within the ACHPR, which is the body that coordinates all human-rights-related issues across the operations of the AU. So far, this body has not yet been erected to the status of a full organ of the AU. In this brief overview of the ACHPR, our interest lies in examining the African approach to self-determination, which is, as intimated earlier, an essential right to the indigenous and local communities and is thus central to the purpose of this work. To do this, we briefly depict how self-determination flows from the Charter of the OAU as it was critical that African countries demonstrate their attachment to this principle during the colonial era. We discuss the current position of Africa in respect of self-determination, as reflected through the deliberations of the ACHPR in the build-up to the conclusion and adoption of the 2007 UN-DRIP.

The naissance of the OAU in 1963 was timely because it was precisely around this period that most African countries were attaining independence. With the establishment of the OAU, African leaders accepted the boundaries between countries as erected during the colonial era and committed to respect them by specifically abiding by the principle of non-violability of boundaries.

Article II of the OAU Charter sets five objectives, including:

1. To promote unity and solidarity of the African states.
2. To coordinate and intensify their cooperation and efforts to achieve a better life for the peoples of Africa.
3. To defend their sovereignty, their territorial integrity and independence.
4. To eradicate all forms of colonialism.
5. To promote international cooperation, having due regard to the Charter of the UN and the Universal Declaration of Human Rights.

To complement these objectives and to provide the basis for cooperation amongst African countries in the realization of the above-mentioned objectives, Article III of the OAU Charter lay down the principles that will guide such cooperation, which are:

1. Sovereign equality of all member states.
2. Non-interference in the internal affairs of states.
3. Respect for the sovereignty and territorial integrity of each state and for its inalienable right to independent existence.
4. Peaceful settlement of disputes by negotiation, mediation, conciliation or arbitration.
5. Unreserved condemnation in all its forms of political assassination as well as subversive activities on the part of neighbouring states.
6. Absolute dedication to the total emancipation of the African territories which are still dependent.
7. Affirmation of a policy of non-alignment with regard to all blocks.

As it transpires from these objectives and principles, decolonization, sovereignty and equality were among the key matters that African heads of states founding members of the OAU, aimed at realizing through the Charter. The inclusion of the principle of self-determination in the ACHPR was therefore well calculated and can arguably be seen as being initially aimed at assisting African countries, in the spirit of cooperation and solidarity, to push forward the process of decolonization. In the early incarnation of the ACHPR, an introductory statement to the meeting of the drafting committee, which was held in Dakar, Senegal, from 28 November to 8 December 1979, it was clearly stipulated that one of the cornerstones of the African Charter on Human Rights was that, in the course of its development, drafters must lay emphasis to addressing the substance of Article II of the OAU Charter. More specifically, this introductory document to the November–December 1979 meeting of the drafting committee stated that African countries had the duty of solidarity and cooperation on state sovereignty and the struggle against colonial domination. The principle of self-determination contemplated in the African Charter was therefore going to be instrumental in assisting African countries to pursue the process of decolonization. However, if decolonization was an important objective of the OAU Charter, the Constitutive

Act of the AU did not include it among its objectives in Article II. That notwithstanding, like the OAU Charter, the Constitutive Act of the AU still stressed the commitment of African countries to the realization of human rights by endorsing the African Charter on Human Rights as its main instrument. But while the principle of self-determination has not been withdrawn from the African Charter of Human and Peoples' Rights, the Constitutive Act of the AU adopted in Lome, Togo, on 11 July 2000 did not mention liberation or decolonization of African countries among its rather expanded list of objectives. It is worth stating that the Charter of OAU included six objectives while the Constitutive Act of the African Union contains 14 objectives, none of which accommodates the pursuit of decolonization.[6] This shows that from an African perspective, self-determination must be given another meaning because liberation from colonial power is no longer the priority within the framework of cooperation among African states. As we discuss in turn, the African approach to self-determination is very relevant to the circumstances of indigenous and local communities, as its inclusion in regulatory measures aimed at the regulation of access to and use of biological resources and TK will empower the communities to control these processes.

Despite lacking the status of full organ of the AU, the ACHPR can be described as the 'ears and voice' of the AU when it comes to human-rights issues. The opinions and views of this body on specific human-rights matters must therefore be considered as reflecting the position of the AU. In respect of indigenous peoples' issues and the right to self-determination, the AU's position on the self-determination principle is among other issues that delayed the adoption of the UN Declaration on the Rights of Indigenous Peoples from November 2006 to September 2007. When the Human Rights Council, the successor of the UN Human Rights Commission, adopted the Draft Declaration in June 2006, it was widely expected that the next General Assembly of the UN and third committee of the Human Rights Council would finally take up the draft declaration to make it the UN Declaration on the Rights of Indigenous Peoples. But this expectation was not met because, while African countries largely voted for the adoption of the draft declaration by the Human Rights Council in June 2006, in November of the same year, Namibia, on behalf of Africa, requested and obtained an amendment that meant the postponement of the adoption of the draft declaration by the General Assembly of the UN. According to Namibia, the declaration is of such importance that there is no need to rush its adoption. Instead, it is critical to give time for more consultations with the view to achieving consensus.[7] One issue for which consultations and discussions are needed is the concept of self-determination as encapsulated by the draft declaration, which, according to Rwanda, was not favouring the integration of indigenous peoples into the wider societies. According to Rwanda, if the draft declaration was adopted by the UN General Assembly as it stood in June 2006, the resulting declaration would establish divisive policies and set a bad precedent. It would isolate groups and incite them to establish their own institutions alongside central existing ones. That would weaken states and hinder their recovery processes.[8]

In order to clarify its position on contentious issues such as the concept of self-determination before its countries can undertake to sign up to the declaration, the ACHPR requested the Working Group of Experts on Indigenous Populations/Communities to provide advice on self-determination. The African perspective on issues such as self-determination was therefore summarized in the Advisory Opinion on the Draft Declaration by the ACHPR, which was issued in May 2007.[9] It was the view of the ACHPR as portrayed in the Advisory Opinion that the right to self-determination, as crafted in the draft declaration, may be understood as the granting of a unilateral right to self-determination and the possibility of cessation to a given section of the population, all such actions representing a threat to the political unity and territorial integrity of a country.[10] This position of the ACHPR is not surprising anyway because the risk of cessation and territorial disintegration have always been associated with the realization of the right to self-determination. The ACHPR stresses that, like all articles of the draft declaration, the article on self-determination must be read and interpreted alongside Article 46 which provides that, 'nothing in this Declaration may be interpreted as implying for any State, people, group or person any right to engage in any activity or to perform any act contrary to the Charter of the United Nations'.[11]

The ACHPR emphasizes that self-determination can be exercised only in the context of Article 46 of the Declaration, which is in conformity with the African Commission's jurisprudence on the promotion and protection of the rights of indigenous populations, based on respect of sovereignty, the inviolability of the borders acquired at the independence of the member states, and respect for their territorial integrity.[12]

It appears that the ACHPR subordinates the realization of the right to self-determination of indigenous populations and communities to a prior respect of political unity and territorial integrity of a country. Therefore, in view to promoting the realization of their right to self-determination over the management of their assets, it is reasonable that local and indigenous communities are empowered to provide their views on how their assets may be accessed and used and the kinds of compensation that they can expect from these operations. To impact on access procedures, these views need to be taken into consideration as key elements guiding the determination of the approval or refusal of the access authorization permits that are administered by the designated national authorities.

Community rights in the Asia Pacific region: the Indian case

In pursuing its objective concerning the promotion and observance of the fundamental human rights and freedoms enshrined in the UDHR and other relevant international covenants, the UN uses regional cooperation among states as an important tool. In this spirit, the UN, through the Office of the Human Rights Commissioner, organizes regional programmes aimed at devising methods for the establishment of regional cooperation on human-rights issues. These types of meetings have been organized in the Asian Pacific region. From the early 1980s until 2004, 12 meetings were organized in various cities of the Asian Pacific countries, starting from 1982 with the first regional workshop held in Colombo (Sri Lanka) to the 12th Workshop on Regional Cooperation for the Promotion and Protection of Human Rights in the Asian Pacific Region, held in Doha Qatar in March 2004. It must be emphasized that these meetings were not discussing community rights or indigenous peoples' rights and concerns in particular: they addressed wide-ranging issues pertaining to the promotion and realization of human rights in the Asia Pacific region, national capacity-building, human-rights education and institutions, regional cooperation and the possibility for the establishment of a regional human-rights agreement or treaty.[13] One of the conclusions arrived at in the course of this process was the consensus that resulted from the Amman Workshop in 1997, which clearly pointed to the fact that, 'the sharing of information and the development and strengthening of national capacities for the promotion and protection of human rights are crucial in the step-by-step process towards establishing regional human-rights arrangements'.[14]

This workshop also recommended the development and implementation of a technical cooperation programme as a matter of priority.[15] At the following workshop, held in Tehran in 1998, the then UN High Commissioner on Human Rights, Mary Robinson, in her introductory address, acknowledged the difficulties the region had been facing in its attempts to establish regional agreements on human rights but stressed that these difficulties should not prevent progress towards this aim.[16] The workshop in Tehran also adopted the technical programme developed by the Office of the High Commissioner which was going to be implemented as a mechanism aimed at achieving the materialization of the regional cooperation on human-rights issues. Progress on the realization of key areas of the technical programme was reported at the 11th workshop in Islamabad in 2003, for instance in respect of the development of national human-rights plans of action and national capacity-building. On this, it must be said that the Islamabad meeting noted the efforts of some countries while encouraging others to follow suit.[17]

It is an unfortunate remark of this work that the development of national action plans on human rights in the region is rather slow. Another critical step achieved within the framework of this regional process was the adoption in Doha in 2004, of the 2004–6 Doha Programme of Action for the Asia Pacific Framework on Regional Cooperation for the Promotion and Protection of Human Rights in the Asian-Pacific Region. This programme of action revolved around four areas:

1. National human-rights plans of action and national capacity building.
2. Human-rights education.
3. National human-rights institutions.
4. The realization of the right to development and economic, social and cultural rights.

As signalled earlier, although these processes make reference to human-rights concepts such as the right to development and economic, social and cultural rights which are central to the concerns of traditional and forest-dwelling communities, they do not mention indigenous peoples' rights specifically. Unlike

Africa, with its charter on human rights and people's rights and its commission on human rights which has set up a working group on indigenous populations and communities, there is thus far no region-wide instrument on human rights in the Asian-Pacific region. Being part of this region, the way India treats the rights and interests of traditional and farming communities must be explored from specific pieces of legislations starting with the relevant provisions of the 1950 Constitution of India.[18]

In India, the official name used to identify indigenous peoples by the government is 'scheduled tribes' (STs). The 1950 Constitution of India provides for the classification of STs and for the management of their lands and resources thereon. This approach by the government to classify indigenous peoples is however not the one by which indigenous peoples identify or recognize themselves. Fundamentally, indigenous peoples in India want to be recognized as the first inhabitants or primary settlers of the areas they inhabit, which is dissimilar to the approach adopted by the government by which the latter can effectively classify and declassify certain communities living in some areas from the category of STs. Based on the government approach to classification, there is an estimated 84 million members of STs in India, representing 8 per cent of the total Indian population. Members of these tribes can be clustered in 461 recognized groups. Based on the number only, members of the STs can be categorized as minorities in India and can therefore benefit from full implementation of the provisions on minority rights of the 1950 Constitution of India.[19] The concept of 'culture' in the provisions on minority rights in the Indian Constitution can obviously be interpreted in a broader sense to include issues such as local communities' traditional use and management of their biological resources and TK which are often sought by outsiders.

However, another characterization of the STs is that they are classified as tribes located in specific states in the north-eastern region of India. Members of these tribes are essentially poor – among the poorest populations in India – and rely heavily on the forest and its resources for their medicinal, fodder, food and energy needs. Based on this characterization, which is linked to the livelihoods of STs, it is our view that there are sections of the Indian population not classified as STs which can be found in many other rural areas in India, pursuing the same livelihood activities as members of classified STs. In this respect, in the context of India, this work does not therefore target the STs as they are classified by the government, but is very much aimed at all communities who live in rural areas and use forest and natural resources as their bread and butter; in other words, they actively pursue the same livelihood activities as members of the classified ST. The development of regulatory measures accommodating such concepts as self-determination, participation in decision-making processes, and the need for PIC on issues affecting them, should be aimed at and involve these communities, not just members of the classified STs. Taking our broadened approach in defining traditional communities who are the targets of the discussions and policy measures contemplated in this work, the provisions of the Indian constitution on STs[20] should not just focus on those STs classified by the government and located in the north-eastern region of India but should also cover all traditional communities and populations living in rural areas whose livelihood is based on natural resources and thus require some legal space that will promote respect of their traditional custodianship of the resources found on and accessed from their traditional lands.

There is an important piece of legislation that has dealt with the rights of traditional communities in a manner that, on the one hand, is aimed at addressing the concerns and interests of members of STs as crafted in the Constitution of India and, on the other, covered the peoples that we target in this work, notably, local and indigenous communities living in forested villages and relying on the forests and its resources for their livelihoods. The Scheduled Tribes and Other Traditional Forest Dwellers (Recognition of Forest Rights) Act was enacted in 2006 by the Indian parliament and received the assent of the President of India on 29 December 2006 (hereafter Indian Forest Rights Act 2006).

The purpose of the Indian Forest Rights Act 2006 is to recognize and vest the forest rights and occupation in forest land in forest-dwelling STs and other traditional forest-dwellers who have been residing in such forests for generations but whose rights could not be recorded. Furthermore, it provides a framework for recording the forest rights and the nature of evidence related to such forest rights. The forest rights that need to be recognized and vested in STs and other forest-dwelling communities are listed in Section 3 of the Act and include, but are not limited to, such rights as the right to hold and live on the forest land under the individual or common occupation for habitation, or for self-cultivation for livelihood by a member(s) of a forest-dwelling STs and other traditional forest-dwellers on all forest lands; the right of ownership, access to collect, use and dispose of minor forest produce which has been traditionally collected within and outside the village boundaries; the right to protect, regenerate or conserve or manage any community forest resource which they have been traditionally protecting and conserving for sustainable use; and the right of access to biodiversity and community rights to IP and TK related to biodiversity and cultural diversity.[21] Section 4 of the Act makes provisions concerning the recognition, restoration and vesting of the rights stipulated in Section 3. In essence, the Act provides that subject to any other law, the government should recognize and ensure protection of the rights stipulated in Section 3 for all members of STs and any other traditional forest-dwellers. Pursuant to the provisions of Section 6 of the Act, the Gram Sabha[22] is the local authority that has the responsibility to initiate the process for determining the nature and extent of the individual and community forest rights or both, that may be recognized in forest-dwelling STs and other traditional forest-dwellers within the limits of its jurisdiction under the Act by receiving claims, verifying and consolidating them. Through a process that entails an opportunity given to members of the community to petition a resolution of the Gram Sabha in respect of the recognition of their forest rights, the rights to be vested by the government to those who are entitled will eventually be based on the recommendation of the Gram Sabha. Finally, there are provisions on offences and penalties stipulated in Sections 7 and 8 of the Forest Rights Act. Sections 7 of the Act provides that any authority, officer, or committee found to have contravened any provision of the Act shall be deemed guilty of offence and punished with fines which may extend to thousands of rupees. However, the second paragraph of this article seems to weaken the content of the first as it contends that someone may rightly be able to prove that the alleged act of offence might have been committed without one's knowledge or that all due diligence was exercised to prevent the commission of such offence. Furthermore, embedded in Section 8 of the Act, there is an added layer of bureaucratic procedure with regards to any pursuit of a legal challenge against anyone who might have contravened the recognized forest rights of scheduled tribes and other traditional forest-dwellers. This section makes the State Level Monitoring Committee an intermediary accessor of grievances and complaints of the Gram Sabha and any other stakeholder, before the court can hear and deal with them. This makes the judicial process complicated, especially for members of the STs and traditional forest-dwellers, if they were able and wanted to pursue court proceedings against anyone who might have infringed their rights. Overall, the Forest Rights Act of India addresses the rights and concerns of indigenous and traditional communities that are discussed in this work. They include ownership rights, which should imply the rights to control access over the assets that the communities own on the basis of their customary systems. In this respect, local communities should fully participate in the policy-making processes pertaining to the regulation of their assets.

Human rights and indigenous peoples' rights in the Andean Community and the OAS: an eye on Brazil and Peru

National approaches for the promotion and realization of human rights in Brazil and Peru are aligned to the policies of two important regional groupings and their integrated human-rights-oriented regulatory instruments. These two groupings are the Andean Community and the Organization of American States (OAS). It must, however, be underlined that while Brazil and Peru are part of the OAS and are therefore expected to comply with their policies and regulations, only Peru is a member of the Andean Community and is thus legally bound by the decisions of either the Andean Council of (Foreign) Ministers or the Commission of the Andean Community.

Known today as the Andean Community, which was created in 1997 in replacement of the Andean Pact, the Pact that created the original Andean Group (the Cartagena Protocol) was signed on 26 May 1969 by five South American countries; namely, Bolivia, Chile, Colombia, Ecuador and Peru. Through this pact, the five countries engaged in a process of regional integration among themselves on various issues, including economic, social and developmental, in order to improve the living standards of their populations. As it stands today, Venezuela is member of the Andean Community as it decided to join in February 1973 while Chile decided to withdraw from the Andean Community in October 1976. The objectives of the Andean Community as stipulated in Article 1 of the Cartagena Protocol include the promotion of balanced and harmonious development of the member countries under equitable conditions to integration and economic and social cooperation; the acceleration of growth and rate of creation of employment, the facilitation of countries' participation in the regional integration process. One avenue contemplated for the realization of the objective of integration and cooperation was via mechanisms provided for in Article 3 of the Protocol, notably the gradual harmonization of economic and social policies and national laws with regards to pertinent matters. Key institutions established under the protocol are entrusted with the implementation of the mechanism of policy integration including the Andean Council of Ministers[23] and the Commission of the Andean Community.[24] Community policies developed by these institutions are published either as declarations, which are non-binding policies, or as decisions, which are the regulatory instruments that are legally binding and must be implemented in the same manner in all the five member countries of the integration system. It is expected that the regulations produced by the community must assist member countries in their drive to attaining, in a cooperative manner, some clearly stipulated social-development objectives such as the full participation by the sub-region's inhabitants in the integration process, paying attention to the needs of the predominantly rural depressed areas.[25]

Human-rights-related policies are key components of the policy actions that the community must undertake on the path of realization of the social development within the framework of the integration approach contemplated by the Protocol. However, the development of human-rights instruments has evolved at a slow pace in the community; the beginning of this process can be traced back as very recently as the 2001 Machu Picchu Declaration on Democracy, the Rights of Indigenous Peoples and the War against Poverty, issued by the Community Head States at a meeting held in Lima, Peru.[26] Through this declaration, the heads of states of the community declared their commitment to 'move ahead with the defense of democracy and respect for human rights'.[27]

In respect of the rights of indigenous peoples, this declaration underlined the firm support of the heads of states to all efforts aimed at the promotion and protection of the basic rights and freedoms of the indigenous peoples, and requested the Council of Ministers to establish a working group on the rights of indigenous peoples whose role will be instrumental in all efforts aimed at the promotion and protection of the rights of indigenous peoples. The wish and request expressed by the heads of states of the community through the Machu Picchu Declaration was met in 2002 with the creation by the Council of Ministers of a working group on the rights of indigenous peoples. Decision 524 of the Andean Community provides that, among other responsibilities, this working group will recommend measures for promoting the rights of indigenous peoples, particularly in terms of the eradication of poverty, development of social equity and recognition of the native contribution to the society of the Andean countries,[28] and give its opinion to the Andean Council of Foreign Ministers, the Commission or the Andean Community General Secretariat, as appropriate, on its own initiative or at their requests, on matters connected with the participation of indigenous peoples in the sub-regional integration process.[29] Membership to the working group is stipulated under Article 4 of Decision 524 and is expected to draw members from representatives of indigenous peoples and organizations, civil-society organizations, government officials and other advisers and observers. However, another significant move by the community was achieved on 26 September 2007, with the adoption by the Council of Ministers, of Decision 674, which was intended to speed up and materialize the implementation of Decision 524 but which, in effect, repealed it.[30] Decision 674 created the Consultative Council of the Andean Community Indigenous Peoples. The composition of the Consultative Council is exclusively made up of representatives from indigenous or national indigenous associations, while some selected regional indigenous organizations will sit in the meetings of the Consultative Council as observers.[31] Among the functions of the Consultative Council stipulated in Article 3 of Decision 674, is its participation in the meetings of the Council of Foreign Ministers and the Andean Community Commission with the right to vote.

Another very important milestone arrived at by the Andean Community in respect of human-rights protection was the development of the Charter for the Promotion and Protection of Human Rights in July 2002. Like the African Charter, the Andean Community followed the approach of developing a comprehensive human-rights instrument, covering various aspects of human rights. Therefore, rather than developing many pieces of human-rights instruments perhaps split to reflect the known generations of human rights,[32] the Andean Charter covers in one instrument several human-rights-related issues, including human rights and democracy,[33] civil and political rights,[34] economic, social and cultural rights,[35] the right to development,[36] and the rights of indigenous peoples and communities of African descent.[37] In particular, Section VIII of the Andean Charter on the Rights of Indigenous peoples and Communities of African Descent has gone to the extent of declaring its commitment to the provisions of ILO Convention No. 169, which is not largely followed by countries hosting large areas of rainforest and large populations of forest-dwelling communities.[38] The core concerns of traditional communities (see Chapter 1) that this work discusses and aims to suggest human-rights-oriented policy and regulatory measures including issues of ownership[39] of land and the resources thereon, collective rights, the protection of collective knowledge[40] associated with the biological resources and concerns over benefit sharing[41] and compensation to local communities are covered by the Community Charter on Human Rights. At least with regards to Peru, as member of the Andean Community, there is strong legal foundation at the community level for this country to ensure that human-rights provisions, in particular in respect of the concerns of traditional communities, are integrated in biodiversity and IP-based policy-making. But Peru is also a member of the OAS, which has its own set-up of a human-rights approach. We now turn to briefly describe that set-up, stressing how it impacts on the rights of traditional and indigenous communities.

Alongside, but larger than, the Andean Community, the OAS is an important regional umbrella organization whose policy and organizational instruments are even referred to in the instruments of the Andean Community. Importantly, Brazil and Peru are members of the OAS, meaning that the human-rights orientations contemplated by the OAS in its instruments are to be complied with equally by these two countries. This work does not intend to embark into a detailed exploration and discussion of the broader institutional framework of the OAS, neither does it intend to examine all the human-rights facets of the OAS. However, what this section aims to do is to very succinctly browse through the human-rights instruments of the OAS and the human-rights provisions of the OAS Charter, looking at the extent to which they can be viewed as addressing, or can be interpreted vis-á-vis possible application to address, the concerns of rural communities. Here, it is critical for the reader to be reminded of our definition of the peoples who are targeted in this work and the issues over which this work's proposed regulatory measures will aim to address. In effect, unlike the work of Thornberry, which focused on indigenous peoples as they are understood to be the primary inhabitants of a given area, the peoples targeted by this work are all those who live in remote forested areas and whose livelihoods are heavily reliant on forest resources. Their traditional farming practices and innovative techniques in breeding local germplasms and cultivating the germplasms themselves, and their knowledge related to the use of their medicinal plants, for instance, are all sought, accessed and exploited by outsiders for scientific and commercially motivated purposes. In this work, we identify them indifferently as traditional societies, farming communities, traditional communities and indigenous communities. The human-rights provisions of the OAS policy instruments are therefore explored with regards to how they may be interpreted to be applicable in furthering the rights-related concerns of these communities.

To the extent that human-rights instruments are applicable to addressing the rights-related concerns of traditional and local communities, the landscape of the human-rights instruments of the OAS comprises:

- The Charter that established the OAS at the Ninth International Conference of American States, held in Bogotá in 1948 (it must be stressed that this charter entered into force in 1951 and has undergone several amendments thus far).
- The 1948 American Declaration on the Rights and Duties of Man and the Inter-American Charter of Social Guarantees.
- The 1969 Inter-American Convention on Human Rights.
- The proposed 1997 American Declaration on the Rights of Indigenous Peoples.

The above-mentioned instruments cover many fields of human rights, some of which are very peculiar to the OAS system. For instance, in its preamble, the OAS Charter declared that the true significance of solidarity and good neighbourliness among member states is through the consolidation within a framework of democratic institutions of a system of individual liberty and social justice based on respect of the essential rights of man.[42] Despite failing to mention the rights of indigenous peoples per se but mentioning issues such as the need for the cultural values of American countries to be cherished and the need for member states to consider themselves individually and jointly bound to preserve and enrich the cultural heritage of the American peoples, two key provisions of the OAS charter can be interpreted to be very relevant to traditional and indigenous communities. These are Article 31, in which the American states proclaim the fundamental rights of the individual without distinction as to race, nationality, creed or sex, and Article 17, in which American states stress that, from their perspective, development of cultural, political and economic life requires respect for the rights of the individual and the principles of universal morality. Yes, of course, the OAS charter may be challenged on the basis that it stresses individualistic rights; however the key elements of this clause which are 'respect for rights and principles of morality' are the elements that must be kept in mind in the process of designing regulatory measures that can address the individual and collective concerns of traditional societies over access to and use of their assets by outsiders.

The American Declaration on the Rights and Duties of Man is divided into two chapters, one dealing with the 'rights' dimension and the other dealing with the 'duties' dimension. Despite stressing individual rights and not mentioning indigenous communities, the declaration covers many territories of human rights among which four are of particular relevance for the purpose of this work as they can be interpreted to be applicable in furtherance of the rights-related concerns of traditional and local communities. These are:

- Article XIII, which covers the right to the benefits of culture.
- Article XIV on the rights to work and to fair remuneration.
- Article XXIII on the right to property.
- Article XXVIII on the scope of the rights of man.

It is indeed our contention that, together, Articles XIII and XIV address benefit-sharing issues by rights and the protection of the IPRs of individuals. This argument stems from the fact that Article XIII encompasses terms such as participation in the 'benefits that result from intellectual progress, especially scientific discoveries' and the notion of 'authorship' which is a conditionality for someone to be entitled the right to protection of moral and material interests associated with their invention. Likewise, based on the stipulation of the second paragraph of Article XIV, traditional and local communities could well argue that they deserve appropriate compensation from the use of knowledge, innovation and practices associated with medicinal plants or the improvement of local germplasm because, for instance, these materials, when accessed and acquired by outsiders, are already in an advance stage of development resulting from the hard labour of local and indigenous communities. Article XXIII on the private right over property represents the basis for local communities to defend their ownership right over their traditionally owned lands. Such traditional lands fit the qualification of private property under this article because it is an important asset that helps local communities to meet the essential needs for a decent living and is central in maintaining communities' dignity. In its approach to delineating the boundaries of the rights of man, the Declaration has sought to ensure that the enjoyment of the rights of one individual does not affect the rights of another. This balance is embedded in Article XXVIII of the Declaration.[43] For instance, based on this article, one can argue that while local communities are entitled to proper compensation from modern exploitation of their innovation and practices, they too need to understand that these benefits cannot preclude scientific and commercial users from benefiting from any further improvement they might have added to the assets collected from local communities. One potentially significant constraint to the enjoyment of the rights as discussed above caught our attention in the chapter on the duties of man. This constraint is encrusted in Article XXXIII of the declaration on the duties to obey the law. Under this provision, every person does not just have the duty to obey the law, but they must also obey every legitimate command of the authorities. The second part of this article is potentially dangerous for local communities who will be expected to obey the commands of local authorities, few of which might be legal or in their interests.

The Inter-American Convention on Human Rights, which was adopted at the Specialized Conference on Human Rights in San Jose, Costa Rica, in 1969, entered into force in 1978.[44] As a product of the works of the Inter-American Commission on Human Rights, which must reflect the spirit and the letter of both the OAS charter and the Inter-American Declaration on Human Rights, this particular instrument does not mention terms such as 'indigenous' or 'local communities'. The text of the convention is also very extensive, as it covers the traditional civil and political rights while briefly touching upon the economic, social and cultural rights under the title 'progressive development'. What caught our attention in the various provisions of the Convention is its stand on issues of participation to public affairs, equality and non-discrimination, as well as the enforcement of the rights through the normal judicial process. The provisions on the protection of human rights over these issues run through Articles 23–25.[45] These provisions are potentially very useful to traditional and local communities. For one thing, because the process through which access permits are issued can be described as a public affair, traditional communities must be entitled, by rights, to participate in this process, based on the fact that they are the traditional custodians of the biological resources occurring on their lands. It should therefore be the case that no law discriminates them from participating in the process through which permit applications are evaluated by the relevant national bodies and eventually granted to outsiders. Furthermore, local communities must have the legal entitlement and right to seek legal remedies through the judicial systems against any operator or entity that infringes their rights over their assets and must thus benefit from the state support as contemplated under Article 25(2).

Of all these instruments, the 1997 Proposed American Declaration on the Rights of Indigenous Peoples is the one that is specifically designed to address the concerns and interests of indigenous peoples in the OAS system. Stretching back from the early 1980s when the OAS recognized the need to develop a human-rights instrument specializing on the rights of indigenous peoples, the Inter-American Commission on Human Rights (IACHR) commenced consultations 'Concerning the Future Inter-American Legal Instrument on Indigenous Rights' in 1992–3.[46] Under the control of the IACHR, a draft of the Inter-American Declaration on the Rights of Indigenous Peoples was approved by the IACHR in

1995. This draft was subsequently revised and approved by the IACHR in 1997 as 'Proposed American Declaration on the Rights of Indigenous Peoples', which was subsequently sent to the General Assembly of the OAS held in Lima, Peru in 1997.[47] From this time onward, work and progress on the proposed declaration has benefited from the involvement of the Inter-American Juridical Committee and is currently pursued and administered by the Committee on Juridical Affairs of the Permanent Council. In 1999, the General Assembly of OAS decided to establish a working group (under the Committee on Juridical Affairs) to focus on furthering work on the Proposed Declaration.[48] There have so far been several resolutions approved by the Committee of Juridical Affairs, such resolutions typically reflecting progress of the working group on the proposed declaration. In line with the general belief that treaty negotiations can be a lengthy and rocky exercise, it appears that until 2008 there was really no sign that the working group would complete its work any time soon, and there is no specific time frame as might have been instructed by the OAS General Assembly for such completion. Indeed, a draft resolution on the draft American Declaration on the Rights of Indigenous Peoples approved by the Committee on Juridical Affairs on 14 May 2008 is illustrative of the lack of a decisive move towards the finalization of the proposed declaration.[49] Section 1 of this resolution is still making the rhetorical affirmation that the adoption of this declaration on the rights of indigenous peoples is a priority to the OAS.[50] In Section 2 of this resolution, the mandate of the working group to prepare the American Declaration on the Rights of Indigenous Peoples is renewed, without any specific timetable set for the completion of its work; while Section 3 of this draft resolution goes on to contemplate the number of meetings that the working group is expected to hold between 2008 and 2009 while stressing the importance of indigenous peoples' participation in these meetings.[51] But this section has failed to at least provide an indicative list of outcomes that should be expected from these gatherings.

Despite the above-mentioned concerns, which could lead to a dangerously elastic process towards the completion of the proposed American declaration, from the perspective of someone who is sympathetic to the concerns of traditional and local communities identified in this work, it is nonetheless a good thing for the OAS to have engaged in this process. The proposed declaration incorporates some provisions which caught our attention in respect of their potential applicability to addressing the concerns of local communities in forested areas. First, in a significant departure from the Declaration and the IACHR, which both stressed individual rights, discussions on the proposed declaration on the rights of indigenous peoples are showing growing consideration of the collective nature of the rights of indigenous peoples.[52] As argued by Patrick Thornberry, the proposed American declaration stresses very much individual rights, despite its recognition of the collective well-being of indigenous peoples as it relates to their traditional systems for use and control of land use.[53] Aspects of the human-rights-related concerns to indigenous communities covered in the proposed declaration extend to the eradication of poverty; the right to development; indigenous culture; ecology; harmonious relations; respect and absence of discrimination; indigenous lands; the right to belong to indigenous populations; rejection of forced assimilation; special guarantees against discrimination; cultural development; the right to environmental protection; organizational and political rights; and the right to self-government (bearing the notion of self-determination). The need to seek and acquire the PIC of indigenous peoples on many issues affecting their lives is recurrently reiterated throughout the highly bracketed text of the 2008 version of the proposed American declaration on the rights of indigenous peoples. In addition, the notion of indigenous participation in decision-making processes, for instance on plans related to the conservation of their lands or the extraction of any resources from these lands, is reiterated in the text of the proposed declaration. Furthermore, the right to legal recognition of indigenous ownership of their lands in a collective or individual manner is covered by the text under discussion as well as the recognition of indigenous IPRs and rights to benefit from its use. Overall, despite the immense presence of brackets in the current text of the proposed declaration, and despite its shortcomings, the text incorporates important provisions that can be potentially very useful in terms of addressing the human-rights-related concerns of local and traditional societies identified in this work.

The 2007 UN-DRIP

The UN-DRIP was adopted on 13 September 2007 under Resolution 61/295 of the 61st General Assembly of the UN. The adoption of this declaration is the result of 25 years of protracted debates among member states of the UN.[54] Important milestones in this process are the adoption in 1994 of the Draft Declaration on the Rights of Indigenous Peoples, which was then forwarded to the UN Commission on Human Rights for consideration and further development. In June 2006, the successor of the UN Commission on Human Rights, the UN Human Rights Council, adopted a further improved version of the draft declaration, which was highly expected to be adopted by the UN General Assembly in November 2006. However, following the submission of requests for amendments by the African group at the meeting of the UN Human Rights Council in November 2006, there emerged a broader consensus that consideration for the adoption of the UN Draft Declaration by the General Assembly should be delayed, to allow more discussions, exchanges and the development of a common understanding on the sticking points raised by the African group. The request for amendments of the African group was submitted by Namibia on behalf of the group. In essence, the African group needed more discussions and clarifications that would contextualize certain issues in the African context including the definition of indigenous peoples, self-determination, ownership of land and resources, the establishment of distinct political and economic institutions and the issue of national and territorial integrity.[55] Negotiations on these issues proceeded through 2007, eventually leading to the adoption of the Declaration in September 2007 by the UN General Assembly.

The 2007 UN Declaration on the Rights of Indigenous Peoples is just a declaration, which, in UN practices, is the sort of legal instrument that can only be used occasionally.[56] However, still in the light of UN practices, the fact that a declaration can be used, albeit occasionally, creates the expectation that members of the international community will abide by it. Thereafter, as its application is gradually justified, a declaration, even if it is a non-legally binding one, can turn out to be largely accepted as the instrument that lays the foundation of binding rules on certain issues (in this case, the rights of indigenous peoples) to countries that have signed up to it. Specifically on the 2007 UN Declaration, the language used on some of its provisions looks as if the declaration was mandatory to countries. For instance, Article 38 of the UN Declaration provides that 'States in consultation and cooperation with indigenous peoples 'shall' take the appropriate measures, including legislative measures to achieve the ends of this Declaration.'[57]

The tone of this provision, strengthened by the use of the term 'shall' sounds as if the declaration is obliging states to consult and cooperate with indigenous peoples in any actions affecting them, because, should they fail to do so, they could theoretically be held accountable of any complaints emerging from indigenous peoples or their representatives. For example, one of the core issues of concerns addressed in this work relates to the participation of traditional communities, by rights, in the decision-making processes pertaining to the issuance of research and plant-collection permits. When such authorizations are granted for research that will involve local populations' TK, innovations and practices, or granted to authorize plant collection that is planned to take place on the lands of local populations, these populations are rightfully entitled to be consulted and to participate in the relevant processes.[58] Based on Article 38, states are expected to enshrine this right in domestic law, in order to ensure the realization of the UN Declaration. But the non-legally binding nature of the Declaration – despite its rather seemingly mandatory language in Article 38 – makes difficult the pursuit of legal actions against or for any legal action to have a significant impact or influence on states that on the one hand do not enshrine local populations' right to participation in domestic law or rather violate this specific right if it was protected by domestic law.

The 2007 UN Declaration is very broad in its coverage of issues of concerns to indigenous peoples. It covers:

- The rights of indigenous peoples including the right to life.[59]
- The right to maintain and strengthen distinct political, legal, economic, social and cultural institutions.[60]
- The right to education and, in particular, the right to establish, own and tailor educational systems to the conditions and needs of indigenous peoples.[61]
- The rights to traditional medicine and to maintain own health practices.
- The right to maintain and strengthen a distinctive spiritual relationship with the traditionally owned or otherwise occupied or used lands.[62]

In respect of the issues addressed in this work, apart from the right to participation discussed in the previous paragraph, there are other provisions of the declaration that are of significant interest for our analyses. These include Article 28 on the right to redress. This Article can be interpreted to be applicable towards the realization of the right of traditional communities to benefit or to be fairly compensated from the utilization of their resources. The provision of such compensation must therefore be based on the recognition that indigenous peoples have rights to their TK, innovations and practices,[63] which should be taken as the IP of indigenous peoples and local communities. Furthermore, the UN Declaration stresses upon the need to consult and cooperate[64] with indigenous peoples and to acquire their PIC[65] in various issues affecting them. An interesting feature running through the text of the UN Declaration on the Rights of Indigenous Peoples is that it recognizes both the individual and collective rights of members of indigenous communities. In translating this to the context of forest-dwelling communities, one can reasonably argue that members of local communities are entitled to own legal rights to control access to their lands, provided they can demonstrate within the boundaries established by any national legal system that they or their ancestors have been owning and using such lands and the resources thereon. On the enforcement of indigenous peoples' rights in cases they are infringed, the declaration appears to be very soft and perhaps unclear. Article 40 vaguely stipulates that indigenous peoples have the right to effective remedies for all infringements of their individual and collective rights. Unlike the other provisions of the declaration, this article does not stipulate the actions that states 'shall' take in dealing with this issue. Nonetheless, the declaration contends that any action to be pursued in this course must take into account the traditions and customs of the peoples affected.

The UN Declaration on the Rights of Indigenous Peoples is a powerful international instrument, which, if contextualized to domestic circumstances, will go a very long way in addressing the concerns of local communities in respect of issues such as bio-piracy and lack of compensation from the utilization of their assets by outsiders in scientific and commercially motivated undertakings.

Conclusion

Discussions carried out in this chapter demonstrate that international as well as regional human-rights instruments bear several provisions that can be crafted in domestic law to address the human-rights concerns of traditional communities in respect of access to and exploitation of their assets in scientific and commercially motivated undertakings. For the purpose of their implementation at the domestic level, national legislators must ensure that they understand the special circumstances of local communities before shaping domestic laws which are aimed at their issues of concerns. For instance, in the case of Brazil and Peru, which are both members of OAS, national legislators in these countries need to be aware that the human-rights instruments of the OAS system are broadly geared towards the individualistic nature of human rights. Knowing that local communities, like any other peoples, have concerns that may be individual or collective in nature, law-makers in Peru and Brazil need to bear this kind of constraint in mind when establishing national legislations reflecting the human-rights provisions of the OAS system which best address the concerns of local communities. Law-making at the domestic level should be able to achieve this, provided indeed that its development process ensures the full participation of the peoples concerned through their elected representatives, which is by itself a way of respecting the rights of local communities.

5 Incursion in the 'bio-piracy' debate

Modern exploitation of the biodiversity components of developing countries and community rights

Against the determinants of bio-piracy discussed in the first chapter, this chapter intends to analyse some allegedly claimed cases of bio-piracy in each of our study countries. The objectives of this exercise are at least threefold:

1. Analysing these cases of bio-piracy will help to ascertain whether they can be characterized and recognized in the determinants described earlier in the introductory chapter.
2. Through this analysis, we will be able to assess the extent to which these cases have had any impacts on the rights and concerns of rural communities discussed herein.
3. This exercise is also going to mark our contribution to increasing awareness about practices leading to what characterizes bio-piracy. From this contribution, stakeholders interested in one way or another in bio-piracy issues will be equipped with the tools that will enable them to make informed decisions in the interests of local communities.

From a strategic perspective, in order to arrive at any conclusion in respect of bio-piracy issues, we must first understand what the people or entities (culprits) generally associated with accusations of bio-piracy actually do, notably plant breeders and researchers or industries involved in the discovery and development of new pharmaceutical drugs. As it emerges from the following, these actors may well appear to hold strong arguments to justify their activities and the profits that they derive from exploiting natural products in association with or based on – arguably – the TK and practices of local and indigenous communities.

Synopsis on plant breeding

In the past, farming and crop selection used to be concomitantly conducted by farmers. To have the materials for their next planting season, the farmers' approach was to select what they saw as the best seeds of their harvests from the current planting season. This approach to crop selection was based on the phenotypic characteristics, which can be simply described as the external features of the seeds. Among the features that guided the selection process, one can mention the shape of the seeds, the seeds' compatibility with post-harvest operations such as drying, as well as the adaptability of the products derived from the cultivation of the seeds, such as fruits and vegetables, to food-processing operations which were of interest to the food industry. For example, when varieties of tomatoes were cultivated for food processing, seed selection would stress such traits as high solid content and thick skins, which could endure mechanical handling during harvesting and transportation to the processing units. Gradually, as farmers wanted to increase their market share, farming activities became separate from seed selection, a process that began in Europe and North America in the late 19th century[1] and which is still ongoing in most developing countries today. Farmers' seed selection, however, was very slow and quite unpredictable because it was, as indicated, based on observation and thus very empirical. This situation changed with the discovery of the Mendelian principles of heredity and their application to seed technology and crop improvement. Based on the Mendelian principles, scientists understood that in actual fact the observable and external (phenotypic) characteristics of a seed, or of the plants and crops grown from these seeds, derive from the non-observable characteristics which are located in the genes (genotypic makeup) of the seeds.[2] With this knowledge, they started seed and crop improvement programmes which could then target specific genes that are responsible for certain characteristics of some crops, and devised methods to transfer them to other seeds that may have other useful genes responsible for other characteristics that are lacking to the other seeds. The gradual mastering and application of the Mendelian principles of genetic inheritance led to targeted development of 'pure lines' of inbred or self-pollinating crops, which are bred-true and uniform crops, made up of traits that can be transferred to other crops. The development of pure lines based on the principles of genetics made seed improvement quicker and more predictable. The innovative forms of seed selection and crop improvement, which evolved from the beginning of the 20th century, are known as conventional plant breeding. Even today, conventional plant breeding still evolves around three major steps:

1. The multiplication of plants having the desirable traits.
2. Assessment of the crops and the identification and selection of individuals that best exhibit the desired traits.
3. The recombination of these quality individuals to generate a new population for subsequent cycles of selection and improvement.[3]

Conventional breeding has many facets which are all applicable to plants depending on whether they are self-pollinating or cross-pollinating. Furthermore, after the main improvements have been carried out and the seeds with the desirable traits obtained, they can be multiplied very quickly using vegetative techniques such as cuttings, grafts or tubers. The basic approach in crop improvement used by plant breeders is known as 'crossing and selection'. To pursue this selection scheme, plant breeders first define the objectives of their selection programme. The objectives are generally aimed at the development of seeds against specifically defined criteria such as yield, resistance to pest and diseases, adaptation to extreme environmental conditions such as drought and winter and seeds' adaptability to storage conditions. Also and importantly, selection criteria are aimed at meeting the expectations of consumers regarding the consumption characteristics of produce that will derive from such seeds and the expectations of the food industry regarding the processing characteristics of the produce. Having defined the objectives of its selection programme, the breeder subsequently embarks on a series of crossing and selection of materials that best fit within its objectives. Two initial parental lines, which, it must be stressed, are already the result of many generations of selection activities undertaken by traditional farmers or professional seed improvers, are crossed, the products of which are field-tested and selected, then crossed again. This process of crossing and selection is carried out in several generations until the breeder is satisfied that the products best fit the traits that are sought. This basic method works well when most of the desirable traits are dominants in the successive lines obtained after crossing and selection. However, there are situations when the desirable traits are promoted by recessive genes in the parental lines which may well be additionally loaded with many undesirable traits. In this case, generally the intention of the breeder is to transfer just the desirable traits to the offspring, leaving the undesirable traits with the parental lines. To achieve this, the breeder applies the technique known as 'backcrossing'. With backcross breeding, the elite parental line having the one and only desirable trait is crossed with a recurrent recipient that may have many desirable characteristics but lack the one and only desirable trait that is found in the other elite parental line. What happens in backcross selection is that one of the progeny from every crossing generation contains the desirable trait of the parental elite. It is always selected and crossed with the other parental line in order to gradually eliminate the undesirable traits of the elite parent while integrating the many desirable characteristics of the original recipient parent. Ultimately, the resulting products will incorporate the desirable trait of the elite parent and all the desirable characteristics of the initial recipient parent.

It is generally understood that the above-described approaches are well suited for self-pollinating crops such as wheat, rice and sorghum, but not for open and cross-pollinating plants such as maize. To address the constraints associated with applying these techniques to open pollinating plants, scientists managed to induce in these plants the characteristic of what they called 'heterosis' as a result of cross-breeding their pure lines.[4] The characteristic of 'heterosis', otherwise known as the hybrid vigour, is manifested by elevated yields, albeit just for one generation. This is advantageous to seed producers because, as the hybrid vigour degenerates in subsequent generations, farmers are obliged to buy seeds for every planting season if they want to maintain similar yields. Additional advantages of hybrids, this time mostly for farmers, is that crops generated from hybrid seeds are uniform. Agricultural mechanization is therefore made easier for farmers during harvesting.[5]

Alongside the conventional breeding techniques, although not in replacement of them, are the biotechnology-based breeding schemes. Indeed, to arrive at the desired seeds, crossing and selecting techniques can be time-consuming and are all based on field-testing and phenotypic and visual evaluation. However, new techniques of genetic engineering and others such as marker assisted selection (MAS) using DNA markers can help to transfer the desired genes from one plant to another even from one crop to a different crop very quickly. These techniques can also help to identify plants which incorporate the desired genes among large populations of plants, thus reducing the length of breeding time in conventional breeding. All these techniques, conventional and biotechnology-based, were applied in the breeding programme of Quality Protein Maize (QPM) pursued by the International Maize and Wheat Improvement Center (CIMMYT) in Mexico. Using a combination of these techniques, the centre succeeded in breeding maize seeds incorporating the genes known as *opaque2,* coding for two essential amino acids in maize, notably lysine and tryptophan. A recessive mutation of this gene was discovered by scientists at Purdue University back in 1963.[6] Looking at the CIMMYT breeding programme on QPM, it is plausible to state that biotechnology-based breeding techniques do not replace the conventional breeding approaches; they complement them and make breeding programmes quicker and more predictable.

Materials used by breeders as initial elites in their seeds-improvement programmes are generally the result of many generations of breeding schemes[7] starting from local farmers who might have been based in a remote village somewhere in the world, all the way through other breeding professionals. These initial elites are not the primitive landraces of the first farming communities that used them in the very first place. By the time plant breeders arrive at the results that address the objectives of their own breeding programme, they might have gone through eight or more generations of crossing and selection and have made substantial investments in terms of time, finances, infrastructures and human resources among others. The products or seeds resulting from their breeding programme are those that will undergo the process of application and protection through PBRs provided they meet the basic criteria of novelty, distinctness, uniformity and stability discussed in Chapter 2. It therefore looks as if arguing for benefit-sharing arrangements and any other impacts of these PBRs over some local farming communities is not an easy task, because such communities may be difficult to identify or may even be unidentifiable. Furthermore, the value added to the PBRs protected seeds make them legal properties of plant breeders because they are the people who shaped them according to the PBRs criteria. It therefore seems logical that discussions concerning misappropriation of communities' landraces that are allegedly protected by PBRs take into account the scientific and technological efforts that plant breeders invest in their breeding programmes. A basic understanding of the type of materials used by breeders and the manner and the improvement level at which breeders acquire their parental lines might mitigate feelings of misappropriation and bio-piracy on the part of local farmers and other actors who advocate the protection of farmers' rights.

Brief on drug R&D

The process of research and development (R&D) leading to the production and, eventually, the commercialization of new drugs is a laborious undertaking. It is largely portrayed by most experts as a very expensive process that requires huge investments in capital, human resources and technological infrastructure.[8] It is also a very unpredictable process as the regulatory framework which lays down the standards that every single drug to be supplied to the market most comply with (in terms of safety requirements, testing and packaging) is generally very rigorous.[9] The development of new drugs can be based on natural products, which are otherwise known as the molecules extracted from medicinal plants. In some cases, these medicinal plants might have been utilized by many generations of traditional and indigenous populations, whose ancestors developed the critical knowledge and practices associated with the utilization of such plants. However, R&D actors are also thought to have developed new and innovative processes that enable them to synthesize new molecules targeting specific diseases, thus avoiding the need to rely on natural products. On the other hand, the innovative processes are able to equip R&D actors with the capabilities to screen big banks of natural products without relying on the knowledge of the traditional communities that might have supplied the raw materials from which the products were extracted.

There are varying estimates regarding the contribution of PGRs in natural products-based drug development both in respect of the profits allegedly made by pharmaceutical companies and in terms of the proportion of natural products in the final (marketable) drug which is approved by a drug regulatory agency. In general, it is estimated that natural products contribute around 20–50 per cent to the total sales of pharmaceuticals.[10] For example, the cholesterol-lowering drug Zocor, deriving from a fungi genus, contributed to Merck & Co.'s US$6.7 billion annual sales, representing 50 per cent of the company's total annual sales in 2001. Other studies addressed the possible links between natural products and drugs deriving from R&D. An analysis of data related to new drugs approved by the FDA or comparable entities in other countries from 1985 to 1995, focusing on cancer and infectious diseases, has been carried out.[11] The results of this study indicate that 62 per cent of the 87 approved cancer drugs were either of natural origin or were modelled on natural-products parents. Furthermore, 61 per cent of the 299 anti-cancer drugs that entered pre-clinical and clinical development were either of natural origin or were modelled from natural-products parents. A follow-up study, published in 2003, analysed data covering an extended period of 22 years from 1981 to 2002 and assessed another sub-division of compounds related to natural products called the 'natural products mimic'.[12] This study indicates that, in relation to cancer, the percentage of small molecules, new chemical entities that are non-synthetic, has remained at 62 per cent over the whole period. However, in relation to anti-hypertensive drugs, 48 of the 74 formally synthetic drugs can be traced to natural product structures/mimics, while among the 10 anti-migraine drugs, 7 are based on serotonin. The importance of natural products as sources of new molecules or new chemical entities (NCEs) but not as sources or new drugs is therefore demonstrated.

However, to achieve the possible gains mentioned above, research institutions and industries undertake complex and complicated R&D processes which, in the context of the pharmaceutical industry, are time-consuming, risky and expensive, costing about US$400 million to US$800 million to produce a successful commercial drug.[13] As a matter of fact, estimates of the overall costs of drug discovery vary significantly, some analysts situating them anywhere between US$375 million and US$1.7 billion. According to one study, the cost of drug development can hit US$1.7 billion if the costs of failed prospective drugs are factored into the actual costs of discovering, developing and launching any one new drug.[14] In the USA, pharmaceutical R&D processes last between ten and fifteen years, at the end of which the FDA may or may not approve a particular drug for safety reasons. The complexity of targeted diseases requires more time for study and, often, targeted diseases are difficult to evaluate. In terms of risks, in about 5,000–10,000 compounds screened at the initial stages in the R&D processes, only 250 enter preclinical testing, five enter clinical testing and one is approved as a drug by the FDA. Only three out of every ten medicines brought to the market generate sufficient revenue to cover the average out-of-pocket cost of R&D. With all these struggles, some researchers question the concept of bio-piracy as, from their perspective, the related debate tends to disregard the investment in time, money and the resources in natural research products. One prominent African marine natural-products researcher of Rhodes University in South Africa commented on the issue of bio-piracy in the context of South Africa marine natural-products research which was carried out from 2001 to 2005 within the framework of a partnership between Rhodes University and the Californian-based Scripps Research Institute.[15] He commented that, while the R&D process is likely to reveal thousands of 'hits' – natural products that will kill cancer cells or malaria parasite *in vitro* – the most tricky and difficult dimension of the process is the conversion of these 'hits' into genuine lead compounds that can be patented and subsequently deployed for eventual drug development. According to this expert, to be successful, the development dimension of the R&D process must overcome many hurdles including:

- ascertaining in-vivo activity of the hits in animals;
- addressing toxicity issues based on animal testing;
- establishing the structure activity trends;
- establishing the mechanism of action and pharmokinetic profile; and
- establishing the physical properties of the hits and their suitability to biological systems.[16]

There is more and more sophistication of technological processes applied in drug development, the utilization of which, as claimed by some analysts, may lead to suggestions that R&D actors (individuals and institutions) involved in drug development do not even need natural products or the TK of local communities in their operations. Indeed, techniques such as combinatorial chemistry, high through-put screening and laboratories-on-a-chip have, arguably, improved the possibility for the production of new leads in the process of drug development.[17] They do this by either synthesizing new molecules targeted at specific diseases or by screening very quickly thousands of hits that are already stored in huge banks. From our perspective, while these new technological approaches may have increased the speed of production of new leads, they are equally novel tools that can be used to take the best from natural products especially when basic information has already been supplied by local communities.

One very important issue that must be mentioned here is that, despite the continuously growing sophistication of drug R&D protocols, many R&D entities are still heavily involved in bio-prospecting activities in the developing countries. They do this by entering into partnerships with developing countries' governments and institutions. Such partnerships are usually negotiated in very restricted circles. The author of this work is aware of many deals that are being or have been concluded between institutions based in technology-rich countries and their partners based in the developing and African countries. It may be easier to assume that the activities and R&D programmes discussed in these partnerships would be less reliant on natural products precisely because such activities are undertaken by very wealthy organizations that are equipped with all the up-to-date technological infrastructures and that employ well-trained and very qualified personnel. For reasons of confidentiality, these activities cannot be discussed in this work. What emerges here is the fact that R&D institutions are indeed still exploring the wealth of biological diversity and TK associated with the use of medicinal plants by local communities in the developing countries. But the question is, provided some specifically identified leads are directly isolated from known and traditionally used plants, and assuming that these leads are exploited for the production of drugs aimed at specific diseases indicated by some traditional herbalists, are these herbalists rightfully entitled to any benefits made by the companies that invested their resources in the R&D process and in the acquisition of patents over the resulting drugs? The fact of the matter is that only the leads or new molecules, the processes of their development or discovery, and possibly the resulting drugs, can be patented. As such, these items are the legal properties of patent holders (R&D entities and pharmaceutical companies). Nonetheless, based on our determinants and as we discuss below, in these circumstances, we can still talk of bio-piracy, meaning that, if local communities do not benefit in some ways, their rights are affected by these processes.

Selected bio-piracy cases and impacts on community rights

The turmeric case of India

The turmeric plant, scientifically known as *Curcuma longa* of the family of *Zingiberaceaea,* is a native Indian plant. The roots of curcuma are rich in a diversity of active compounds, known as curcuminoides, which are responsible for the multiplicity of potentials of the plant, upon which it has been used from time immemorial in Indian traditional systems. Indeed, when they are transformed into powder, the roots of the turmeric plant exhibit important culinary, cosmetic and medicinal properties.[18] The utilizations of turmeric products as a culinary and a colouring agent are widely known and largely proven scientifically. Furthermore, while the traditional medicinal uses of the turmeric powder are well documented in

Indian texts on Ayurvedic medicine and practised in contemporary traditional therapeutic systems in India for the treatment of ailments such as anaemia, asthma, burns, conjunctivitis, dental problems, diabetes, diarrhoea, and pain, the use of turmeric products to address these health problems is not commonplace in industrial countries. Yet, the alleged case of bio-piracy is essentially associated with the medicinal credentials of turmeric. As we demonstrate below, the key events related to this case have impacted on the rights of local communities, as they are aligned to our description of the determinants of bio-piracy discussed in the first chapter.

Turmeric taken outside India

Turmeric plants are native to southern and south-eastern India. According to a most respected physician, Hakeen Hashmi, the various potentials and properties of turmeric have been mentioned in early writings of the world in many languages.[19] This indicates that the use of turmeric in its various facets that were probably only known to Indian's traditional societies had gone beyond their traditional geographical boundaries and beyond India a long time before any scientific research on turmeric and any patents or any other IPRs were awarded in association with a turmeric-based product. How did the turmeric plants and the information associated with the use of its products leave India and south-east Asia to find themselves in other parts of the world? There may not be a clear-cut answer to this critical question. However, as discussed by other authors, there are many means by which plants and other biological resources have left their traditional areas of origin to be domesticated elsewhere.[20] Perhaps the simple fact that people move from one part of the world to another carrying items of particular usefulness to them can be part of the answer to this question. Another one could be linked with the exploration of useful resources in the colonial territories by scientists from the colonial power. In any case, as far as the turmeric's spread around the world is concerned, it is contended that while it was grown in India during ancient times, it seemed to have reached China before the 7th century AD. The next move was its expansion throughout the West Indies, and it is said to have been transported eastwards across the pacific by Polynesians to Hawaii and Easter Island. With these movements, it appears that, by the time turmeric materials and related information about their utilizations was leaving India and south-east Asia for other parts of the world, there was no issue of 'bio-piracy', and neither were there any issues about regulations or ethics in relation with accessing and exploiting the resources from developing countries. Access was not based upon or did not require any legal authorization or permit as the traditional lands of local communities were declared 'no man's lands' (*terra nullius*).

Patents, the turmeric business and benefit sharing

The very common fashion to present a bio-piracy case is the one which implicates an overseas actor, a researcher, a research institution or a pharmaceutical, cosmetic or seed corporation heavily involved in the search and collection of useful resources from the bio-cultural wealth of countries of the Southern Hemisphere. What is generally described is that these actors, most of them operating through national counterparts, find ways to collect (based on a legal authorization or even illegally) medicinal plants or plants with strong cosmetic potentials and the TK and practices of local communities associated with these plants. With these assets at hand, these actors take them back to their countries where the technology to exploit their full potential is available. There, they transform these assets into other products or processes and seek legal protection of their IPRs by means of for instance patents. What follows is the commercialization of the legally protected outcomes and potentially the making of huge profits which will not be shared with the communities and other actors in the source countries of the raw materials. Our discussion on and presentation of the determinants of bio-piracy have gone beyond this common and rather basic fashion to stylize this issue. Our approach contends that bio-piracy is not just performed by overseas actors but it also very much emerges from the actions of national actors. The turmeric case is perfect evidence. In this case, Indian scientists are among those that are awarded the most contentious patents on turmeric.

In 1995, two Indian scientists, Suma Das and Hari Har Choly, working at the University of Mississippi Medical Center in the USA, were awarded US patent No. 5,401,504 for the use of turmeric powder in the treatment of wounds. The patent document claims a new method of healing wounds in particular skin ulcers through the administration of a wound-healing agent that contains an effective amount of turmeric. The patent goes on to provide details of controlled laboratory experiments performed on rats and other clinical historical cases about how turmeric was used on humans after other therapies had failed. This patent was very controversial because the methods and new processes claimed therein did not appear to be really new, that is, different to existing practices in India and of ancient Indian Ayurvedic medical texts. It was successfully challenged by the New Delhi based Indian Council for Scientific and Industrial Research (CSIR) materialized by the 1996 decision of USPTO that overturned the patent. Of course, to do that, CSIR had spent Rs 500,000 in fees and made use of the experts' legal assistance. It has to be pointed out that this patent is not the only one that is associated with turmeric products. There are indeed hundreds of US patents on turmeric products that could also be challenged on the grounds of lack of novelty, some of which are held by Indian scientists.[21]

There is significant business surrounding turmeric products which are largely sold as dietary supplements and less commercialized as drugs. In fact, turmeric-based pharmaceutical drugs are yet to be approved by regulatory agencies such as the American FDA. Searching online, one finds countless sites offering turmeric-based products at various prices.[22] For example, turmeric products are often presented as anti-ageing agents, helping to maintain a healthy brain and healthy joints and as having the ability to perform anti-Alzheimer's actions.[23] There is a great deal of commercial exploitation of the turmeric products which are largely related to their virtue as a spice. To exploit this facet of the turmeric potentials, companies have been established in India as far back as in the 1950s.

For example, LANCO is an Indian company that was established in 1952 and specializes in the commercialization of Indian turmeric spice and chillies as 'spices of highest international standards' 'to millions of spice loving peoples worldwide'.[24] However, despite the obvious intense business surrounding turmeric products, there is yet a precise figure to be attached to the commercial value of turmeric products worldwide. Any figure is highly likely to amount to millions of US dollars of annual commercial value of turmeric products if all its business-related facets are taken into consideration. It is, however, not clear how local communities whose fathers, forefathers and ancestors have been using turmeric roots from time immemorial, are benefiting directly from these businesses.

Community rights as affected in this case

As it appears, this case reflects the bio-piracy determinants discussed in the first chapter. It is a case that demonstrates how the operations of national actors can drive bio-piracy. In this case, the traditionally known uses of turmeric products are those upon which all the business facets of turmeric are based. Bio-piracy occurs because there is no indication that the commercialization of turmeric products (which are dear and rooted in Indian's traditional way of life) was beforehand authorized by or agreed upon by representatives of Indian traditional societies. Even before the incursion of regulatory and ethical questions in the discourse on access to and use of biological assets and TK, there should have been some form of consultation, at least on moral grounds, between Indian traditional communities and business operators regarding the latter's commercial ambitions on turmeric. Such consultations did not happen. Turmeric-related patents were awarded since 1995, after the entry into force of the CBD and the conclusion of the 1994 WTO-TRIPS Agreement. Of course, discussions on issues of PIC and benefit sharing within the framework of these forums have intensified just recently. Nonetheless, one can note an increase in controversial turmeric-related patent claims over processes similar to traditionally known uses of turmeric products, such claims that could have failed the test of novelty. Understandably, these patents may not be the main drivers of the businesses surrounding turmeric products because owning a patent does not guarantee successful commercialization. But there is no indication that traditional Indian communities are benefiting from all the turmeric-associated businesses, perhaps via a trust fund where the monies from turmeric profits could be put and administered in the way that can make a visible impact to the lives of identifiable communities through the realization of identifiable projects.[25] Community rights to benefit from turmeric products are thus seriously affected by the many businesses that exploit the various facets of turmeric properties. On the other hand, it is our view that, with the inclusion of such principles as PIC of local communities in environmental and human-rights instruments, the lack of compliance with this principle by various actors involved in the turmeric-related operations is also evidence of trespass to community rights to PIC.

The *Pelargonium* case of South Africa

Geographical origin, traditional uses and collection of Pelargonium SP

Two species of the genera Pelargonium are looked into in this case. They are *Pelargonium sidoides* and *Pelargonium reniforme,* both native and endemic plants of southern Africa (South Africa and Lesotho). Botanical records of SANBI suggest that *Pelargonium sidoides* is endemic to South Africa and Lesotho.[26] Its geographical distribution in South Africa extends to the Eastern Cape, the Free State, and the southern and south-western Gauteng. The geographical distribution of *Pelargonium reniforme* in South Africa is not as extensive as the distribution of *Pelargonium sidoides* because *Pelargonium reniforme* is reported to occur mainly in the Eastern Cape and in a limited scope in the Western Cape.

Ethno-botanical records show that traditional practitioners and local communities in South Africa have been using these plants in the treatment of many ailments within the framework of their traditional therapeutic systems. Indeed, as we describe below, when knowledge about the traditional modes of utilization of these plants and the plants themselves moved from local communities' settings to the world stage back in the 1890s, the interests of actors in the developed countries in these plants was directly related to their traditional therapeutic properties. Traditionally, these plants are reported to have been used by many ethnic communities such as the Zulu, Xhosa, Basuto and the Mfengi.[27] These communities have used these plants in the treatment of ailments as varied as respiratory tract and lung infections; tuberculosis; gastrointestinal disorders, including diarrhoea and dysentery; hepatic disorders; wound healing; menstrual pains and many other viral and bacterial infections and conditions such as coughs, fever, fatigue and weakness. If the above-mentioned conditions can be seeing as being typical human conditions, the plants were also applied on various animal and livestock conditions including liver disorder, diarrhoea and dysentery.[28]

The transfer of the traditional medicinal knowledge related to the use of the *Pelargoniums* in the treatment of TB and access to and acquisition of the materials themselves by Western scientists occurred as far back as in 1897. That year, an Englishman Charles Henry Stevens, who was diagnosed with TB in Birmingham in England, travelled to South Africa. On arriving in South Africa, Charles heard that a Zulu traditional doctor based in Lesotho, called Kagaitse, held the expertise over the treatment of TB using local herbs. He went to see this traditional practitioner, who administered him concoctions made from the roots of the *Pelargonium* plants. After three months of treatment, he recovered. Steven returned to England that year with the roots of the *Pelargonium* plants, and began to make and sell concoctions from these roots under the trade name 'Stevens Consumption Cure'. Stevens was also checked by the doctor who had diagnosed his TB a few months earlier and was declared treated. However, despite being successfully cured on the one hand and himself subsequently providing the very same treatment to other sick people in England who then recovered, Stevens Consumption Cure was never authorized by the British

Medical Association (BMA) because this institution did not hold any scientific (in the modern sense) information about the botanical properties of the plants.

As the traditional medicinal knowledge related to these plants and the materials were acquired by the West, scientific research began in the 1930s when a Swiss medical doctor began testing the plants. Over nine years this Swiss medical doctor administered 800 TB patients, recorded his findings and published them in 1930, 1933 and 1937.[29] In a parallel move, the plants were also tested around the same time at Charité-universitätsmedizin Berlin, tests that produced positive results. Based on the successes of these scientific investigations, a company was set up in Germany in 1939, ISO-Arzneimittel, to import the roots of *Pelargonium sidoides* and *Pelargonium reniforme* from South Africa. This represented a significant step towards the huge exploitation and collection of *Pelargonium* roots from South Africa as they are exported to scientific and commercial actors in Germany.

Phytomedicinal exploitation of **Pelargonium** *and actors involved*

ISO-Arzneimittel and the University of Munich have been conducting research to identify the specific characteristics and therapeutic properties of the *Pelargonium* species which, in 2002, led to ISO-Arzneimittel embarking to file international patents on their research outcomes (more on the patents below). These patents and the research endeavours from which they emerged are all very recent, while, as indicated above, the *Pelargonium*-based products were already in the markets as far back as in the late 1890s (e.g., the Stevens Consumption Cure). Nowadays, the Schwabe's Group, a German-based group, is the main and leading international actor in the phyto-pharmaceutical exploitation of the *Pelargonium* plants. The Schwabe Group includes Dr Schwabe Pharmaceuticals, which was founded in 1866 and is currently ranked the 25th largest pharmaceutical manufacturer in Germany, making sales in the region of €420 million in 2004. Furthermore, Dr Schwabe Pharmaceuticals is ranked 149th largest over-the-counter health-care company in the world in terms of market share. It must be stressed that the specific contribution of the Pelargonium-based drugs produced by Dr Schwabe Pharmaceuticals to the performance of the company is not known. Apart from Dr Schwabe's Pharmaceuticals, the other members of the Schwabe Group include its partnerships with actors such as Nature's Way, Madaus AG, Spitzner, Farmasan, Extracta, DHU and ISO-Arzneimittel. The group works with a partner in South Africa, Parceval, based in the Western Cape, which is responsible for ensuring the supply of the raw materials to the Schwabe Group in Germany.

The Schwabe Group manufactures two homoeopathic remedies which are sold by its members as Umckaloabo in Germany and Umcka in the USA. In Germany, Umckaloaba is described as the modern therapy for infections, containing highly concentrated extracts of the active ingredients of the South African *Pelargonium.* The remedy is administered for respiratory tract infections such as bronchitis in children and adults. On the other hand, Umcka is sold in the USA by Nature's Way, a member of the Schwabe Group. This remedy is advertised as being clinically tested to shorten the duration and reduce the severity of sore throats, sinus and bronchial infections. The remedy was put to the US market in 2003 and has witnessed sales increasing to 700 per cent faster than any other drug on the market.[30]

To ensure continuity and sustainability of the R&D (the production of *Pelargonium*-based drugs) within the Schwabe Group, the South African partner of the group, Parceval, is responsible for the supply of raw materials. Parceval supplies materials harvested from its plantations of medicinal plants. But Parceval also buys materials harvested from some areas of the Eastern Cape Province by local harvesters who supply their materials to Parceval through middlemen. It is established that, as compensation deriving from their harvesting endeavours, local harvesters receive US$0.42–2.09 per kilogramme of dried roots. The middlemen who transfer these materials to Parceval in Cape Town earn in the region of US$139 per kilogramme. One critical point to raise is that harvesters appear to be applying unsustainable methods during their harvesting spree such as completely pulling the plants from their roots. These harvesting methods do not allow regrowth of the plants, whose populations have been dwindling significantly from the wild. This depletion of wild populations of *Pelargonium* led the government to issue a moratorium over the wild harvesting of the plants. However, as it appears to be the case, this moratorium has been breached, and the culprits have been sentenced to three months' imprisonment. To address the issue of illegal harvesting, the Department of Environmental Affairs and Tourism issued three collection permits to Parceval and granted the relevant transport permits to ensure that the materials are legally transported from the Eastern Cape to the Western Cape and are legally exported overseas.

Pelargonium-*related patents*

A report by the ACB published in 2007 highlighted the existence of four international patent applications on the Pelargonium, filed through the Patent Cooperation Treaty (PCT) system which is administered by WIPO. However, by the time of writing this part of the book in February 2009, there emerged an additional international Pelargonium-related patent, lifting the number of international patents up to five. One of the five international patents, WO/2008/125239, filed on 4 April 2008, was filed by Dr Willmar Schwabe GmBH & Co. According to the abstract of this patent, the invention in question relates to production methods for obtaining dry extracts from *Pelargonium sidoides* and/or *Pelargonium reniforme,* extracts obtainable according to said methods and pharmaceutical products comprising such extracts.[31] This patent has not yet entered the national phase in any of the designated states. In addition to this patent, three international patents, all filed by ISO-Arzneimittel and subsequently assigned in 2007 to Dr Willmar Schwabe GmbH are WO2005041993, WO2006002837 and WO003028746. These patents have now been granted respectively in Germany for the first (WO2005041993 granted as Patent No. DE10350338 on 7 April 2005) and by the EPO for the second and third (WO2006002837 and WO003028746 respectively granted as Patents No. EP1651244 on 29 August 2007 and EP14297951 on 13

June 2007). These patents raise serious questions in respect of their validity stemming from doubts about the novelty of the claims upon which they were issued. For example, Patent No. WO2005041993, which, it must be stressed, is the most aggressively pursued as it is currently pending in thirteen countries including South Africa,[32] is based on claims that reflect the traditional therapeutic applications of the *Pelargonium* in South Africa. This patent relates to the use of the extracts of the *Pelargonium* plants in the treatment of behavioural changes from disease-related causes, sickness due to stress-related causes and the methods of processing the extracts. Among the examples of symptoms included in the patent are conditions such as fatigue, depression, weakness, anxiety, sleep depravation and anorexia. As obvious as it appears, symptoms such as general fatigue and weakness caused by illness are among the conditions that are treated by traditional doctors when they administer concoctions of *Pelargonium* roots to their patients. Charles Henry Stevens, who was treated in the 1890s, knew this very well as he learnt from being a patient treated by a Zulu traditional doctor. As discussed earlier, modern R&D endeavours incur huge expenses and are very risky and unpredictable activities. Nonetheless, when all the efforts invested in R&D processes end up just validating the traditional ways of using the raw materials, it is hard to distance them from the 'bio-piracy' discourse. What further increases the difficulty of subtracting them from bio-piracy is the fact that there is no involvement of the communities in the processes through consultations which might have sought their consent to engage their assets in scientific and commercially motivated undertakings. In the absence of community engagement, it is not surprising to find no benefit-sharing or compensation scheme between communities and users of their assets. Lack of community participation through consultations and failure for them to be compensated in the case of the *Pelargoniums* make it a bio-piracy issue.

Implications of this case to community rights

In many respects, community rights are affected by the *Pelargonium* case. For one, despite the fact that Parceval is domesticating *Pelargonium* among its arsenal of medicinal plants, the parents of the currently domesticated *Pelargonium* species were once extracted from the wild. They were collected from traditional lands without the consent of traditional authorities. Current collection permits granted to Parceval to collect *Pelargonium* from the wild are also certainly targeting traditional land areas where these plants can still be found naturally. There is no indication that the authorities issuing these authorizations have sought to acquire the consent of local communities in the Eastern Cape, prior to granting these permits. Community customary land rights and self-determination are seriously affected by these proceedings. Furthermore, there is no evidence of benefit sharing or research agreement between say, Parceval and the communities living in the areas where these resources are exploited. Here I cannot even dare to think about the possibility of an agreement between the German-based Schwabe Group and local actors in South Africa. The only benefit that the community may earn from the *Pelargonium*-related business is the payments, or rather compensation, they receive through harvesting *Pelargonium* roots and supplying them to Parceval via middlemen. There is no benefit that they earn on a right-basis from the many business activities that are based on the exploitation of the *Pelargonium* plants.

The roots *Lepidium peruvianum* (maca) from Peru

Geographical origin, traditional uses and movement of TK

Commonly known as Maca in Peru and internationally, *Lepidium peruvianum* is a plant species that originates from the Peruvian highlands. The plant is known for its resistance to extreme conditions such as cold and strong winds, which is why the rare plant was found growing naturally by the first pastoral tribe, the Pumpush, which settled 12,000 feet above Lake Chinchaycocha, today called Lake Junin, in the Peruvian highlands. For a number of reasons, three successive tribes domesticated this species. The primary reason for domesticating this species was because it was found to be most tolerable and adapted to the prevailing extreme climatic conditions. Subsequently and importantly, the need to domesticate Maca became even greater because with its adaptation to climatic conditions, the species was used as staple food by the successive tribes, including the Pumpush, the Yaro and the Inca. Maca has therefore been domesticated and used by traditional Peruvian tribes as far back as 4000 BC. Through use, the prevalent characteristics and potentials of the species were discovered early on, especially in respect of the fertility and aphrodisiac properties of the plants.

With its various potentials, cultivation of and ownership over Maca farms became a highly political issue. This became evident when the Spanish conquered the Inca in the 1500s. Indeed, after the conquests, the Inca offered to their conquerors 10,000–15,000 pounds of ground Maca roots as a precious gift. Even after their conquests of Peru, it was known that the Spanish rulers did not like Peruvian food. However, they just could not do without Maca roots for two main reasons:

1. High altitudes made them (the men) nearly infertile and concoctions from the roots were the remedy.
2. They needed something that could fortify their warriors, and Maca roots were traditionally known to increase strength and fortitude.

In fact, it appears that it was only after fifty full years of occupation that the birth of the first Spanish baby occurred in Peru, thanks to the use and assistance of Maca roots. The roots of Maca were especially shipped to Spain to be used specifically by the royal family. The Spanish conquests of Peru and the shipping of Maca-related knowledge and materials to Spain therefore marked the departure of the assets from Peru. Of course, by then nobody knew what could be the place and role of these Peruvian assets in today's phyto-medicinal R&D and in the market of phyto-medical products used against infertility, impotence and other ailments. Worldwide knowledge about the properties of Maca increased with the publication of various books that proliferated in the 1960s and the 1980s, most of them authored by German and American scientists who were interested in the botanicals of Peru.[33]

From the early crystallization of knowledge about the traditional uses and potential of Maca until today, the TK and practices about the use of Maca have grown and have been passed on by word of mouth through generations. There are many virtues that this species is believed to hold including in relation to its use to increase fertility in humans and livestock alike; its ability to relieve frigidity in women and impotence in men; its adaptogenic virtues as revitalizer of internal organs and as a regulator of menstruation; and as a reliever of symptoms of menopause. It is also recommended for malnutrition, convalescence, memory loss, mental debility, as a general tonic, and in the treatment of respiratory ailments.[34] It is reported that some traditional doctors do not recommend the administration of Maca to people suffering with hypertension.[35] However, regardless of the level of satisfaction and trust that the traditional Peruvian communities hold in the virtues of Maca in treating the above-mentioned ailments, the fact remains traditionally produced medicines from Maca cannot enter the modern market of prescription drugs due to lack of scientific evidence. This is where R&D comes in to take Maca outside the box of traditional therapeutic systems, raising issues and concerns over bio-piracy.

R&D and actors involved

Pure World Inc., now acquired by Naturex, a company based in France, is the major player in the commercial exploitation of the Peruvian Maca and the owner of the controversial patent granted by the US Patent and Trade Mark Office in 2001. Pure World Botanicals, which was wholly owned by Pure World Inc., was the New Jersey-based company that operated the largest extraction facilities of active ingredients from natural plants for the production of nutraceuticals in North America. The company presented itself as specializing in the development, manufacturing and commercialization of natural ingredients which were principally derived from plant materials (also referred to as botanicals or herbs). To undertake these activities, Pure World uses its proprietary extraction, purification and granulation technologies.[36] In 2005, Pure World claimed to have produced more than 1,000 botanical extracts which were used by the cosmetic, food and flavour, nutraceutical and pharmaceutical industries to manufacture finished products for the consumer market.[37] The Peruvian Maca (*Lepidium meyenii* and *Lepidium peruvianum*) is just one of the over 1,000 plants exploited by the company. In a press release issued on 18 July 2005, it was announced that Pure World Inc. had been successfully acquired by Naturex, leading Naturex to anticipate that with this acquisition it had become a major player in the business sector of nutraceuticals. Indeed, since acquiring Pure World in 2005, the revenues of Naturex have grown steadily from €50.25 million in 2005[38] to €70.1 million in 2008.[39] The production and sales of nutraceuticals (not specifically from the

Peruvian Maca, although it plays an important role in Naturex's business in nutraceuticals) is thought to represent as much as 55 per cent of Naturex's consolidated revenues.[40]

On 5 April 2000, Pure World announced in an article published by the peer-reviewed journal *Urology,* its discovery of an extract from the Peruvian Maca, dubbed MacaPure. According to the company, tested on mice and rats, MacaPure has exhibited its properties as a sexual-performance enhancer. The product does this by significantly helping in problems of erectile dysfunction and markedly increasing libido. Pure World Inc. officially hailed its discovery by saying, for instance, that the company arrived at such results by conducting the most comprehensive research ever undertaken on Maca. However, it is well known that Maca has been traditionally used for centuries as a fortifier and sexual enhancer, raising doubt about the novelty of the discovery made by Pure World, supposedly through research. On the other hand, as we discuss below in connection with the patent that was issued over this discovery, it is questionable whether the process through which Pure World arrived at its discovery is non-obvious. Doubts about the novelty and non-obviousness of this patent raise serious questions concerning its validity, pointing to the bio-piracy direction.

Pure World (now Naturex) is not the only actor involved in scientific and commercial exploitation of the Peruvian Maca. As it emerges below, other companies such as Biotics Research Corporation also acquired patents related to Maca. The widespread knowledge of the properties of Maca has obviously led other actors in Japan and Europe to embark on scientific and commercial exploitation of Maca. To this end, they are aided by Peru-based organizations and farmers who increased their cultivation efforts of Maca in order to meet the increasing international demands in raw materials. Farming efforts and supply of raw materials to the international market appear to be the only means through which some fractions of Peruvian communities might be benefiting from the Maca-related businesses. Back in 2002, it was already reported that a Peru-based company, Quimica Suiza, operating as the national subsidiary of the Swiss Pharmaceutical Giant AstraZeneca, had invested US$1 million in Maca R&D.[41] The company worked in collaboration with the International Potato Center (CIP), which maintained large collections of Maca seeds, '*in trust*' for public interests. The company manufactured and exported US$150,000 worth of Maca tablets to Japan, which were promoted as 'natural Viagra'.[42] It is not known whether there is any benefit accruing to the areas where Maca originated.

Maca-related patents

A search of the WIPO PCT patents database releases six international patent applications, some of which claim novel methods of extraction of the active ingredients of *Lepidium meyenii* for use in the treatment of ailments such as skin problems, cancer, sexual dysfunction and infertility.[43] Some of these patents were filed by Pure World Botanicals, such as patent WO/2000/051548, which relates to the 'compositions that can be isolated from Lepidium plant materials and to methods of their isolation. The compositions are useful for treating and preventing cancer and sexual dysfunction.'[44] Among the six international patent filings directly related to *Lepidium meyenii* is the most recent one, Patent No. WO2008012628, filled by CHIEREGATI Angelo, an Italian company, and published on 31 January 2008. This patent relates to a preparation for infertility treatment which includes the following: *Lepidium meyenii,* manganese, vitamin E, selenium and zinc. By January 2009, there was no indication that this patent had entered a national phase. The prior art description of this particular patent is somewhat weird. It vaguely acknowledges that traditional medicine addresses problems of infertility and sexual dysfunction whilst having many side effects, but fails to recognize that Maca has been used in the treatment of sexual disorders for centuries by traditional Peruvian communities, although it points to the fact that its invention relates to the extract of plants known as Maca that grow spontaneously in South America.[45] One can only suspect that failure by this company to be honest on the prior art was meant to protect the supposed novelty of what they claim to be their invention. This failure is undoubtedly intentional and is no more than a legal theft of centuries-old Peruvian traditional medicinal knowledge and practices.

The legal theft of Maca-related medicinal knowledge in the treatment of sexual dysfunction and infertility was first materialized with the acquisition by Pure World Botanicals of US Patent No. 6267995, extract of *Lepidium meyenii* roots for pharmaceutical applications on 31 July 2001. The abstract of this patent describes a process for the isolation and extraction of Maca compositions, using alcohol.[46] It further contends that the isolated composition is aimed at the treatment of cancer and sexual dysfunction.[47] To Dr Qun Yi Zheng, the leader of the team of scientists that undertook this discovery, their intellectual efforts helped them to discover in Maca the compounds that were not known previously. They dubbed these compounds Macaenes and Macamides and tested them to show that they had no side effects. On that basis, Pure World is legally entitled to their patents. The perception of theft of the Peruvian knowledge and practice through this patent is very strong and understandable too. On the one hand, Pure World was very much aware that through their centuries-old use of Maca, traditional communities in Peru have always applied Maca products to the treatment of the very target identified by Pure World. In this respect, it is very surprising that the patent does not even mention the peoples from whom the native knowledge came. On the other hand, if following a scientific method Pure World scientists managed to identify the natural substances of Maca, their process might not really be new and non-obvious in that it mimics the traditional systems of soaking dried roots to separate the active substances, as utilized by traditional Peruvian communities for centuries. According to one expert, Professor Carlos Quiros of the University of California at Davis, the method applied by Pure World scientists was a very standard one. In his opinion, their methods were not different from the traditional Peruvian practices.[48] In this respect, the US patent on Maca held by Pure World is not valid. Unsurprisingly, there is an ongoing international effort aimed at overturning this patent.[49]

Effects on community rights

It is certainly not clear how much revenue the key actors mentioned in this case are earning directly as a contribution of Maca in their overall businesses. Undoubtedly, they gain from Maca whether financially or otherwise. But there is no indication of any benefit-sharing arrangements in place between these actors and some traditional communities in Peru. Perhaps the State of Peru might gain from the locally established corporations that transform Maca roots and export overseas through a taxation system. However, as is well known, the system of tax redistribution and local investment in developing countries – Peru in this case – is not always suited to local development needs. The right of traditional communities to benefit from national and international scientific and commercial exploitation of Maca is therefore affected. Patents related to Maca are filed without taking into account the consent of traditional communities of Peru and without even mentioning their prior use of Maca as prior art. The right to PIC of Peruvian communities for this exploitation of their assets in scientific and commercial motivated undertakings is also infringed. Also, this case demonstrates that national actors contribute to bio-piracy. Blames of theft or failure to share the proceeds from exploitation of the assets of the communities therefore needs not to be directed solely to overseas actors.

The Brazilian cupuaçu

Geographical origin and traditional uses of cupuaçu (Theobroma grandiflorum)

Cupuaçu, also known as *cupuassu* or *copoasu*, is a small to medium tree which grows in the rainforest canopy of the Brazilian Amazon. Scientifically known as *Theobroma grandiflorum*, this plant finds its origin from the south and south-eastern Amazonia in Brazil and is native from the states of Para and Maranhao. Nowadays, the plant can be found in other regions of the Brazilian Amazonia such as in the areas of Tapajos, Xingu and Guama rivers. Indigenous peoples in the rainforest of Amazon are reported to have cultivated and used *cupuaçu* traditionally as food. The fruits of this species are known for their creamy, exotic-tasting pulp, which is used in Brazil and Peru in the production of juice, ice cream, jam and tarts.[50] The seeds were traded in the Rio Negro and Upper Orinoco rivers. In these areas, indigenous peoples drink *cupuaçu* juice in order to facilitate difficult births, but the juice will only be effective as medicine if blessed by the shaman (medicine man) beforehand. Because *Theobroma grandiflorum* is very close to *Theobroma cacao,* it is also used in the production of chocolate-like products. However, there are significant differences between the nutritional properties of *cupuaçu* and *cacao*, which make *cupuaçu* a more attractive raw material for the production of chocolate-like foodstuff. Indeed, cocoa contains the stimulant theobromine, which is a close relative to caffeine. *Cupuaçu* does not contain this particular ingredient, making it a healthier alternative to *Theobroma cacao.* In

Brazil, there is a growing industry for chocolate-like products produced from *cupuaçu*, where they are known as cupulates, while in Japan, 'cupulate' is already produced and commercialized at an industrial scale.[51]

International awareness about the existence, the nutritional properties and the various traditional methods of processing and using *cupuaçu* has grown to the extent that food companies established in Japan and the USA have made its transformation an important dimension of their overall business. To keep such activities going, these companies rely heavily on the supply of raw materials, the fruits of the plants harvested from wild populations in the Brazilian forest. Two points arise here. For one, apart from supplying raw materials of *cupuaçu* fruits through the traditional channels for local exploitation in Brazil, Brazilian suppliers' main role in the international market of *cupuaçu* products is as supplier to the main companies involved. This is, incidentally, the main means by which Brazilian actors benefit from these international activities. Second, heavy extraction of *cupuaçu* for national and international markets threatens wild populations which are increasingly being depleted.

Commercial exploitation and actors involved

Awareness about international exploitation of *cupuaçu* became public when a group of Brazilian NGOs including the NGO Amazon Link (established in September 2001) launched in 2002 a campaign specifically targeted at this plant. The *cupuaçu* campaign, dubbed 'Cupuaçu is Ours', was prompted by the discovery by Amazon Link of the existence of some foreign patents related to cupuaçu, held by a foreign Japanese food corporation Asahi Foods Co. Ltd. Asahi transforms *cupuaçu* fruits for the production of cupulate, a chocolate-like foodstuff which is reported to have been manufactured throughout Brazil since 1983. In order to meet the demands of its domestic food industry, Japan imported more than 50 tons of *cupuaçu*, only in the first quarter of 2002 from the Brazilian state of Amazon. Back in 2002, it was also contemplated that Japan would import more than 200 tons of *cupuaçu* fruits from Brazil in 2003, in order to meet the demand of its cupulate industry.[52] There is as yet no specific figure tagged to the international market value of *cupuaçu* neither for its derivative products such as cupulate. But with the domestic exploitation in Brazil and the international demands such as those from Japan, it will not be a surprise that *cupuaçu* is of significant international market value, thus representing an important earning device to all actors involved in its commercial exploitation. As it appears, both the presumed market value of this plant and the need to provide for an alternative source for the raw materials prompted research initiatives towards its domestication.[53]

Despite its ownership over most of the known patents either granted at some regional or national patent offices (Japan Patent Office [JPO] or European Patent Office [EPO]) or filed through the PCT system at WIPO, Asahi is not the sole international company involved in commercial exploitation and transformation of *cupuaçu*. Other companies such as The Body Shop are exploiting the cosmetic properties of *cupuaçu* and own a patent in relation to it. Also, cupuaçu International Inc., a San Diego-based American company, filed an international patent application with WIPO back in 2002.

Patents and other cupuaçu*-related IPRs*

Despite its exploitation in many regions in Brazil for the production of chocolate-like products, overseas companies such as Asahi have filed for and have even been granted patents over cupulate products and related processes. Obtaining patents over cupulate by a commercial actor is vital and legal evidence that the invention over which the patent is granted derives from creative endeavour of the patent owner. But is that the case with the *cupuaçu*-related patents? What is the current scope of patents and other IP associated with *cupuaçu*?

Online patent searches reveal the existence of many patents filed in major patent offices, most of which are related to some compositions or the process of making such compositions and associated with butter extracted from *cupuaçu*.[54] Generally speaking, these patents claim the scientific methods designed and applied by the inventors through the development of the relevant products.[55] Specifically, Asahi is the owner of two Japanese patents that mention *cupuaçu*. These are JP2001299278 and JP2001348593 respectively, claiming 'Lipids originating from cupuaçu, method of producing the same and use thereof' and 'Oil and fat derived from cupuaçu – Theobroma grandiflorum seed, method for producing the sane and its use.' Apart from its Japanese patents, Asahi owns an EPO patent, EP1219698A1, published in July 2002, and had filed a PCT patent application with the WIPO, No. WO0125377, also published in July 2002. In addition to Asahi, The Body Shop, a UK-based cosmetic corporation, was also identified as owning a UK patent, No. GB2321644A, published in August 1998, over a cosmetic composition comprising *cupuaçu* extract. Furthermore, cupuaçu International Inc., a San Diego-based corporation, filed a PCT patent application, published in November 2002 as No. WO02081606 and covering *cupuaçu* seed-origin fat, the process for producing the same and use thereof.

Another interesting facet of the *cupuaçu* case is that trademarks have been registered in Japan, Europe and the USA in relation to this species for various classes of products including chocolate. In particular, the Japanese trademark No. 4126269 registered by Asahi in 1998 was another critical component of the 'Cupuaçu is Ours' campaign mentioned earlier. Arguably, unlike patents, trade marks do not create individual exclusive rights in the cupuaçu plant itself. However, the owner of the mark can enforce his or her rights by, for instance, preventing others from selling products bearing the protected name. In the case of *cupuaçu*, the Japanese company was indeed able to prohibit other actors, including Brazilian actors who wanted to sell *cupuaçu* products in Japan, from doing so. This situation therefore appeared to have harmed rural producers of *cupuaçu* products in Brazil, who could not use the word '*cupuaçu*' on their products meant for export to the Japanese market.[56] Fortunately, the Japanese patent office annulled this trademark on 1 March 2004, which represented a significant victory for the 'Cupuaçu is Ours' campaign. Perhaps this campaign was successful due to the support it received from a strong coalition of actors including an experienced international law firm Baker Mackenzie, the Brazilian foreign ministry, the Brazilian National Institute of Intellectual Property, the Brazilian Institute of International Trade Law and Development and the University of São Paolo Law School.[57]

Arguably, the above-mentioned patents do not affect the local industrial production systems of *cupuaçu* derivatives in Brazil, and neither do they prevent rural communities from using the plants locally. Furthermore, most of these patents were granted on the grounds that their claims covered scientifically proven 'inventive' processes that are 'industrially applicable', thus being able to lead to the production of products useful to mankind. But perhaps if these patents were thoroughly searched and examined, some of them would not have been granted on grounds of failing to satisfy the novelty criteria. For example, back in 2003, NGOs were already exploring ways to challenge the European patent No. EP1219698A1 because it covered *cupuaçu* oils and chocolate.[58] However, on the economic front these patents could have been beneficial to the source country of the raw materials if, on the one hand, they proved to have contributed to the creation or strengthening of the domestic markets. Another additional benefit to the source country would have been for the patent to demonstrate its contribution in the improvement of in-country technological capacities through mechanisms such as technology transfer. But in the case of *cupuaçu*, such economic benefits to the source country, Brazil, and in particular its rural communities, have not been reported and are highly doubtful.

The real industrial exploitation of the raw materials is carried out overseas including in Japan and the USA where major processing industries are based, leaving the source country – Brazil – to essentially act as supplier of raw materials. In this respect, it is hard to figure out how technology will be effectively transferred in order to enhance source countries' technological capacities. However, the processing industries will find it difficult to convince the large public and rural communities that TK and practices related to the raw assets did not inspire or guide them in the development of their modern and scientifically proven processing methods. In other words, Asahi, for instance, will find it very hard to convince the pubic that the company did not 'pirate' the TK and practices of rural communities that were modernized through scientific and technological means.

Impact of this case on community rights

In many respects, the community rights outlined in the first chapter are impacted in this case. For one, there is no evidence that the right of PIC of Brazilian communities has been sought and acquired within the framework of all the scientific and commercial undertakings involving *cupuaçu* raw and manufactured materials. Second, there is no evidence of the existence of a benefit-sharing agreement between either the national or international actors involved in the exploitation of *cupuaçu* and the communities living in the areas where the plant originates. Furthermore, assuming that harvesting of *cupuaçu* fruits for exportation or domestic transformation is legally authorized by the relevant authorities, there is no indication that local communities are involved or participate in the process through which decisions over such authorizations are made and the authorizations themselves granted. Harvesting is very likely to be conducted on the traditional lands of some communities, even if the latter might not own the legal titles as evidence of land ownership. If access to these land areas is therefore authorized by the relevant authorities to commercial harvesters with no prior consultation between the authorities and traditional peoples, then community land rights are also impacted. Legally speaking, patent holders do not need the consent of local communities when they apply for patents, based on their scientifically researched products and processes. However, in this work, bio-piracy is also rooted on a moral pillar. Thus, morally speaking, a company applying for a patent over *cupuaçu* derivatives is morally wrong if it does not consult communities who have been using this plant for centuries.

Conclusion

This chapter has described what R&D actors do when it comes to adding value to the biological and genetic resources of developing countries. It shows that with all the efforts invested in time, money and human ingenuity, in most cases they are very much entitled to protect their IPRs over the products or processes resulting from their research endeavours. However, in many respects as well, R&D actors do not acquire the raw materials legally or, even when acquired legally, they fail to exploit them in the way that acknowledges prior ownership and use of such materials by local and indigenous communities. Based on such failures, 'bio-piracy' is a reality in their activities and is more so because they exploit community assets without compensating the communities who have been using them since time immemorial or without allowing them to benefit by 'rights'. Apart from the right to benefit, local communities are impacted by bio-piracy cases in many other respects, as they witness the various failures by actors to respect their land rights, to respect their right to participation in decision-making processes, and to respect their right to PIC.

6 Soft and regional undertakings aimed at community rights

Analyses carried out in the previous chapters suggest that based on their customary systems of managing and using their assets, indigenous and local communities have genuine claims of ownership over them. From a human-rights perspective, international and regional human-rights instruments provide for community rights to control access to their assets and to be fully involved in the decision-making processes through which legal access authorizations are granted by the permitting authorities to scientific and commercial operators. Furthermore, human-rights instruments make provisions for local communities to have their free and prior informed consent sought within the framework of the decision-making and other processes pertaining to access to and potential use of their assets. On the other hand, analyses in the previous chapters demonstrate that scientific and commercial bodies have the capacities to undertake commercially oriented research and development exploitation of the assets of local communities once they have acquired them. Legally speaking, having invested time, money and their intellectual ingenuity in the transformation of community assets, it is always highly likely that scientific and commercial users will seek legal protection and deservedly earn IPRs such as patents over the outputs of their undertakings. For local and indigenous communities, knowing that they shared their assets with outsiders freely, based as we said earlier upon the flexibilities of their traditional resource-management systems, the acquisition of patents associated with the likelihood of commercial exploitation of their assets by scientific and commercial users raise concerns such as 'bio-piracy'. Some professional bodies, indigenous peoples associations, international institutions, the research community and countries grouped in regional organizations have developed soft declarations and other legal instruments aimed at community rights in connection with scientific and commercial exploitation of their assets. These instruments attempt to promote respect for the rights and interests of indigenous and local communities over their assets by external users. This chapter has selected and discusses some of these soft measures and legal instruments, stressing their strengths and weaknesses particularly vis-á-vis the protection of community rights and interests.

Strengths and weaknesses of soft declarations over community rights

The first category of instruments developed by indigenous peoples' associations and certain professional bodies are the declarations which generally result from meetings organized by these groups. Through these declarations, coalitions of indigenous peoples' associations express their common views on issues of critical interests to their members in relation to access to and exploitation of their assets by outsiders. A number of such declarations and statements of indigenous peoples' associations have thus been elaborated to promote the legitimate rights of rural communities. In 2002, an assessment of some of these 'voluntary instruments' was carried out by Graham Dutfield, who identified about 15 of such declarations and statements.[1] In his assessment, Dutfield found that these instruments cover a wide range of issues, expressed as the critical demands from rural communities within the framework of access and utilization of their assets. Such critical demands of rural communities include, but are not limited to:

- their ownership and inalienable rights over TK and the biological and genetic materials,
- their PIC;
- their right of full participation to decision-making processes;
- their right of control over access to their assets for the purpose of research, and the right to veto research which is based on their assets including their knowledge, forest resources and their traditional and ancestral lands;
- their right to acquire the full results of research carried out upon their TK and associated biological and genetic materials;
- their rights to receive compensation and their fair share of the benefits arising from the exploitation of their assets.[2]

If these declarations and statements bear significant strengths and legitimacy, it is equally the case that they bear significant weaknesses. On the strengths side, it is noticeable that these declarations are frequently cited by scholars and policy analysts and are therefore well known and represent influential tools used in discussions in the policy circles that address matters related to access to genetic resources, benefit sharing, and community rights. Furthermore, these are declarations and statements which were agreed by representatives of many indigenous peoples' associations across the world. Consequently, they reflect some degree of international consensus. Moreover, these statements and declarations target a large international audience comprising a range of stakeholders having genuine interests in biodiversity regulatory policies including indigenous peoples, national and international policy-makers, civil society organizations and intergovernmental organizations. In this respect, they are carefully drafted and very sophisticated. However, in relation to their weaknesses, these declarations appear to offer no practical guidance to stakeholders, especially scientific and commercial users of the assets of local and indigenous communities, about how to request the PIC of rural communities in the manner that recognizes them as key actors and partners in the decision-making processes pertaining to access. Also, they provide no practical guidance on how rural communities may veto access to the resources or their traditional lands considering that, generally, the legally established permitting authorities grant access authorizations without prior consultation with the communities where collection of plant specimens and ethno-botanical research are to be undertaken. Overall, these declarations and statements have merely identified the above-mentioned critical demands without suggesting practical measures to achieving them. The regulatory measures contemplated in the subsequent chapters are indisputably workable attempts to correct these weaknesses.

Other international attempts and opportunities to address community concerns

Another category of measures developed in an attempt to address the rights and interests of indigenous and local communities in the context of access to and use of their assets in scientific and commercially oriented undertakings are those developed by international professional organizations. These organizations include industries operating in the biology, chemistry and biotechnology sectors that have interests in the exploitation of PGRs. In addition, botanical gardens have also come together to express their common views, crystallized in a set of measures which they believe can go a long way to addressing the legitimate concerns of local and indigenous communities. Thus, some major corporate bodies, including the European Chemistry Industry Council (CEFIC),[3] the biotechnology industry organization (BIO)[4] and the International Chamber of Commerce (ICC),[5] formulated schemes intended to address the interests of rural communities about the scientific and commercial exploitation of their bio-cultural assets. For example, on 10 June 2002, Carl B. Feldbaum, President of the BIO, announced in ten points what was called 'the biotechnology foreign policy', the new path of the biotechnology industry. Points 8 and 9 address bio-prospecting activities and the need for the organization to develop guidelines respecting the cultures of the communities where such activities are taking place, the importance of fair compensation to indigenous peoples and the necessity for the biotechnology industries to promote sustainable development. For its part, the ICC 'supports initiatives aiming at exploring the options for the protection of TK whether within the framework of the existing IPRs regimes or through the development of a *sui generis* regime, which address the rights and interests of indigenous and local communities'.[6]

Also, governments, through the Cusco Declaration[7] of the Like-Minded Megadiverse Countries,[8] on Access to Genetic Resources, Traditional Knowledge and Intellectual Property Rights have demanded the establishment of a patent-based international regime on ABS. It must be stressed that discussions on either the development of options for legal protection of TK and the rights of local and indigenous communities or for the establishment of an international instrument in the form of a *sui generis* system are taking place in international forums. Such forums include the WIPO Intergovernmental Committee on Intellectual Property and Genetic Resources, Traditional Knowledge and Folklore; the WTO-TRIPS Council and the Ad-Hoc Open-Ended Working Group on Access and Benefit Sharing of the CBD. The unfortunate fact arising from these meetings is that there is no indication that all of these debates will soon reach a consensus on, for instance, a legally binding patent-based ABS regime. Nonetheless, it is the view of this work that the lack of international consensus over ways to legally address the rights and legitimate concerns of local and indigenous communities should not discourage biodiversity-rich countries such as Brazil, India, Peru and South Africa to adopt workable measures tailored to their specific conditions.

The 2001 common policy guidelines (CPG) for participating institutions (including botanical gardens, herbaria and museums) on Access to Genetic Resources and Benefit Sharing addresses the concept of 'best practice'.[9] As a concept that can contribute to building trust between users and suppliers of PGRs and TK, 'best practice' is suggested to all those who seek access to and use of these assets. Depending on the national context, whether or not bio-materials are found in *ex-situ* collection, whether or not there is national ABS legislation, seekers for access should endeavour to comply with 'best practice'. Some policy experts define 'best practice' as follows: 'Standard of practice that are widely regarded by those in the field as representing the highest levels of conduct and the practical implementation of core underlying principles such as conservation of biodiversity, sustainable use and equitable benefit sharing among stakeholders.'[10] Perhaps, if scientific and commercial users of the bio-cultural assets of local communities comply with 'best practice' during their activities, rural communities will have the incentive to pursue biodiversity conservation and share their knowledge in the interests of humanity.[11]

TRRs and community concerns

With the inappropriateness of the modern IP regimes (especially the patent regimes) to mitigate the concerns of rural communities about scientific and commercial exploitation of PGRs and TK, there are arguments pointing to the development of national and international *sui generis* biodiversity regulations.[12] International law is supportive for the establishment of such *sui generis* regimes and encourages countries rich in biological and cultural diversity to initiate legislative processes at the national level.[13] The ILO Convention No. 169 Concerning Indigenous and Tribal Peoples in Independents Countries[14] does not specifically use the term '*sui generis*' like Article 27(3)(b) of the WTO-TRIPS Agreement in the context of plant varieties. However, Article 2(1) of ILO No. 169 provides that governments shall have the responsibility for developing, with the participation of the peoples concerned, co-coordinated and systematic action to protect the rights of these peoples and to guarantee respect for their integrity. Furthermore, Article 8(1) of ILO No. 169 stipulates that in applying national laws and regulations to the peoples concerned, their customs and customary laws shall be duly taken into consideration. The CBD provides for the respect, preservation and maintenance of knowledge, innovations and practices of indigenous and local communities, subject to national legislation.[15]

Countries rich in biological and cultural diversity can take advantage of these opportunities under international law to establish enforceable instruments at the national level addressing the concerns of rural communities. In this respect, the process for the design and establishment of such a national instrument can be carried out under the guidance of the TRRs concept,[16] taking into account the way this concept has defined (see the section on the theoretical foundation of national ABS measures below) and accommodated the assets of traditional and local communities.

Genesis of the TRRs concept

From a historical perspective, the early incarnations of this concept can be traced back from the First International Congress of Ethnobiology in 1998 in Belem, Brazil. At this conference, indigenous peoples from various parts of the world met with scientists and environmentalists to discuss a common strategy to stop the rapid loss of the planet's biological and cultural diversity. Item 4 of the declaration of Belem was the first time an international scientific organization recognized a basic obligation that procedures should be developed to compensate native peoples for the utilization of their knowledge and their biological resources. The second World Congress of the Society of Ethnobiology held in Kunning China in 1990 yielded an action plan called the Kunning Action Plan. This action plan led to the creation of the Global Coalition for Biological and Cultural Diversity, which brought together indigenous peoples, scientific organizations and environmental groups in a working group on IPRs. The objective of this working group was to develop a strategy for the use of TK, the involvement of local peoples in conservation and development. In 1994, the Working Group on Intellectual Property Rights changed its name to the Working Group on Traditional, Intellectual, Cultural and Scientific Resources Rights. The rationale behind this change of name was to try to mitigate any idea of closeness between this working group and the profit-making drive spirit of the conventional legal framework of IPRs while adopting and adapting existing hard and soft law to the specific needs of indigenous and local communities embodying traditional lifestyles.[17]

TRRs: theoretical foundation of national biodiversity measures

Within the framework of the TRRs concept, traditional resources include both tangible and intangible resources valuable to mankind, including their spiritual, aesthetic, cultural and economic values. These resources are used by rural communities on a daily basis and allow them to maintain the pride of unique heritage and healthy environment. The approach of TRRs in the protection of the rights of traditional and indigenous communities over traditional resources is based on the Four Directions Council's definition of the word 'traditional'.[18] According to the Four Directions Council, the word 'traditional' is not synonymous with antiquity, which undermines the values of the related resources. Rather, 'traditional' should be understood as indicating that the value associated with the resources has emerged from an accumulation of innovative efforts of local communities over time.

The TRRs concept suggests developing countries to explore the possibility of using four pillars in their efforts to develop national biodiversity regulations around which they can tailor their biodiversity regulatory measures at the national level. These pillars are:

1. human rights (both communal and individual rights) such as the right to development, the right to self-determination, land and territorial rights;
2. a great deal of soft laws that can be considered in the process. These soft laws include declarations,[19] codes of ethics[20] and treaties;
3. harmonization of the policy engagement of government departments vis-á-vis international agreements and declarations to which the country is a member party;
4. national regulatory systems that ensure equity through the *equitizing* process.

In other words, financial means and other opportunities should also be given to indigenous peoples enabling them to take part in international discussions where issues impacting their life are debated. They should also be empowered to undertake local development activities.

There are fears and doubts regarding the appropriateness of the TRRs to properly regulate biodiversity activities and to satisfy the interests of rural communities. Despite the fact that TRRs stress cultural and spiritual values of traditional resources to rural communities, this concept does not seem to be different from the commercial and utilitarian features of the modern IP regimes, especially patent regimes.[21] Shane Greene argues that, in order to address the concerns of rural communities about access to and scientific or commercial use of PGRs and TK, they must be given greater power to control access at the local level. The whole concept of TRRs should therefore be narrowed down to securing land and territorial rights to indigenous peoples because only such rights will provide them with real empowerment.[22]

This work acknowledges that the principles of territorial/land rights and self-determination are embedded in the TRRs concept. However, it is, from a human-rights perspective, their practical implementation in terms of ensuring the full participation of rural communities in the decision-making processes that have not yet been tested in a specific context. The regulatory measures proposed in the subsequent chapters represent an attempt to correct this weakness and by so doing will test the practical dimension of the TRRs.

The African Model Law for the Protection of the Rights of Local Communities, Farmers and Breeders and for the Regulation of Access to Biological Resources

The African Model Law for the Protection of the Rights of Local Communities, Farmers and Breeders and for the Regulation of Access to Biological Resources is the outcome of an involved process of research and consultations coordinated by the Scientific, Technical and Research Commission (STRC) of the predecessor of the AU, the OAU.[23] Understandably, because the process was sponsored by the government of Ethiopia, key technical players who partnered with the STRC were the Ethiopian Environmental Protection Agency and the Institute for Sustainable Development of Ethiopia. The whole process started with the organization of a workshop in 1997 in Nairobi, Kenya, by the STRC on 'Medicinal Plants and Herbal Medicine in Africa: Policy Issues on Ownership, Access and Conservation', purposely to address what the STRC identified as a problem of control, conservation and utilization of Africa's biological resources. The process culminated with a draft model law tabled for discussion at the 68th Ordinary Session of the Council of Ministers of the OAU, which was held in Ouagadougou in Burkina Faso in June 1998. The OAU Council of Ministers expressed their very strong support to the draft model law by adopting it and recommending that OAU member states do the same and initiate a process at the national level involving all stakeholders to enact it into law or use it as guide in the development of a domestic law. The implication for just recommending member countries to adopt the model law at the national level is that, as it stands, the model law is not a legally binding instrument to members of the AU. However, AU members are encouraged to ensure that their national legislative processes on ABS reflect very much the spirit and content of the model law. The objective of the African Model Law is to: 'give reasoned attention to agricultural development (food crops and medicinal plants), indigenous knowledge systems, conservation and sustainable use of biological resources (natural forest products, fish, animals, bacteria and other), community rights, equitable sharing of benefits and national sovereignty consistent with the imperatives of the Convention on Biological Diversity'.[24]

The African Model Law attempts to address the issue of legal recognition of the customary rights of local communities as 'community rights' and has devoted an entire section to this. Despite the similarities in the wording of many articles of the CBD and the African Model Law, the latter has gone a step further by elaborating on the rights of rural communities that should be recognized and legally protected. Article 16 of the Model Law expands on the assets over which the rights of local communities should be recognized and protected by states, including their ownership rights over biological resources, innovations and practices; their right to collectively benefit from the use of their biological resources, innovations and practices; and their rights to use their innovations, practices, knowledge and technologies in the conservation and sustainable management of biological diversity. Article 17 goes a step further to stress that in providing for the recognition and protection of the rights identified in Article 16, such recognition and protection must be pursued in a way that such rights are enshrined and protected under the norms, practices and customary law found in and recognized by the concerned local communities whether those norms and laws are written or not. In addition to the provisions of Articles 16 and 17, other critical rights of rural communities covered by the African Model Law and which are the centre piece of this book, include their right to:

- be consulted within the framework of the PIC process;[25]
- refuse consent over access to their resources, their technologies and practices;[26]
- withdraw or place restrictions on consent and access;[27]
- traditional access, use and exchange of the materials;
- benefit from the utilization of their resources, technologies and knowledge.[28]

To access and acquire the assets of local communities, actors (organizations and individuals) seeking them employ a variety of strategies. Some actors follow the national legal process, meaning that they thrive to comply with the national permitting system. In respect of this kind of process, overseas institutions are generally required to establish partnerships with national stakeholders with whom they will collaborate in the course of the research process. However, national academic and research institutions can access and acquire the assets of local communities without any specific legal authorization other than their status as public-research institutions that supposedly undertake research for the public good. Also, eco-tourism is another means by which individuals collect traditionally valuable medicinal plants and get information informally from passive villagers, who will be sharing them with the tourist not knowing that the latter has genuine but hidden plans about how to exploit such information. Whatever means is applied by scientific and commercial actors or any individual to access the assets of the communities, it is generally the case that those means are disrespectful to local systems of resource management, take advantage of the flexibilities of such systems and undermine the rights of rural communities. What is meant by flexibilities of local systems of resource management is the fact that, in principle, rural communities share their knowledge among themselves and with outsiders freely and allow access to the resources on their lands, not necessarily for profit-making purposes but because such resources can be useful to others and to humanity. However, the increased awareness of rural communities about the scientific and commercial potential of their resources and their mounting concerns about the lack of fair compensation they receive from scientific and commercial users of their resources are making them turn to secrecy. Arguably, a careful accommodation by national patent-based biodiversity regulatory measures of issues along the lines of those considered by Article 11 of the African Model Law, which relate to conditions pertaining to academic and research institutions, public agencies and the intergovernmental institutions in respect of access to community assets, will improve rural communities' control over access in terms of ensuring their participation in the decision-making processes.

An important issue that has to be considered in relation to scientific and commercial exploitation of PGRs and TK is the application of patent systems for the protection of products or processes derived from such activities. The African Model Law covers this issue both in its provisions on the suggested elements worth considering in agreements[29] and those addressing patents over life forms and biological processes. The approach of the Model Law is the one of a total ban of patent over life forms.[30] From my perspective, such a position is inappropriate for the technology-poor African countries because it may obstruct the establishment of profitable partnerships between national and overseas partners over the exploitation of valuable biological resources of the country. Instead, a more realistic and practical approach to patenting can be devised, aiming at protecting local innovations, providing benefits to local stakeholders and contributing to technology transfer. All these issues must be based on the prior recognition and respect of the human rights of local communities to PIC; participation in the decision-making processes through which legal authorizations are granted to individuals; and to scientific and commercial actors to exploit community assets.

Decision 391 of the Andean Community

It should be recalled that an important dimension of the objective of the Andean Community, as crafted in the 1969 Cartagena Protocol that created the initial Andean Pact (see Chapter 4 above for further information on the evolvement of institutional and integration process of the Andean Community), is the promotion of conditions of regional integration as well as economic and social cooperation among the member countries of the Andean Community. The community contemplates the realization of this element of its objectives through the establishment of harmonized laws and policies in such forms as 'decisions', which can be adopted by the Commission of the Cartagena Agreement and hold the status of legally binding instruments to all member countries of the community. It is within this framework that on 2 July 1996, after its adoption by the Commission of the Cartagena Agreement, member countries of the Andean Community signed Decision 391 which established a common regime on access to genetic resources.

In many respects, Decision 391 is very relevant to this work. This decision covers the assets of traditional and local communities over which there is need for the development of workable regulatory measures for the protection of community rights and addressing community concerns from a human-rights perspective. Like the previous instruments examined in this chapter, Decision 391 has a number of strong points, for instance in respect of its recognition of the value of genetic resources and the intangible TK and practices of local communities. Decision 391 describes these elements as 'research and economic assets' and as 'valuable cultural, spiritual and livelihood assets' to local and indigenous communities who are identified in the decision as native, Afro-American and local communities.[31] Other strong points of Decision 391 include its approach in regulating access through the acquisition of the legal authorization dubbed as 'Access Resolution',[32] which is an order issued by the relevant competent authority that executes administrative procedures pertaining access to genetic resources and its by-products,[33] and the signing of access contracts which must include benefit-sharing provisions. Importantly, Article 7 of Decision 391 which relates to the recognition of the know-how, innovations and traditional practices of local communities, request member countries to recognize and value the rights and the authority of the native, Afro-American and local communities to decide about their know-how, innovations and traditional practices associated with the genetic resources and their by-products. To ensure that Decision 391 has teeth, meaning that it is enforceable, it provides for sanctions to those who dare to undertake access to and exploitation of the genetic resources and by-products in contravention with its provisions. Indeed, Article 46 of Decision 391 stipulates that anyone undertaking access and exploitation without the required 'Access Resolution' and with no contract should be punished. Article 47 of Decision 391 entrusts the relevant competent national authority with the responsibility to pursue enforcement procedures, empowering it to apply for administrative sanctions and possibly the disqualification of the violator, pursuant to the procedure in force in the member country. The provisions for access and the sanctions associated with their violation are all meant to achieve the purpose and aims of Decision 391 which include, among other issues, the establishment of conditions for just and equitable participation in the benefits of the access[34] and the laying of foundations for the recognition and valuation of the genetic resources and their by-products and of their associated intangible components, especially when native, Afro-American or local communities are involved.[35]

From the perspective of this work, the weak points of Decision 391 stem from its approach to achieve the dimension of its aims identified in Article 2(b), especially in respect of encapsulating the concerns of local communities from a human-rights perspective. Many human-rights-related concerns of local communities, as we contemplate them in this work, are very poorly covered by Decision 391. For one thing, despite its recognition and request to member countries to value the rights of local communities to decide upon their assets (as stipulated in Article 7), the detailed access procedure provided under Articles 16 and 17 of Decision 391 have surprisingly failed to accommodate local communities' rights to participation in the procedure of access. As they stand, the provisions on the procedure through which access contracts are signed and eventually the 'Access Resolution' issued do not require any consultations with local communities and fail to require their free and prior informed consent. The provisions of Decision 391 on the Access Contract simply require actors involved in contract negotiations to 'bear in mind' the rights and interests of the suppliers of the genetic resources.[36] This approach to addressing the rights and interests of the suppliers of the resources fail to identify local communities, who are the traditional custodians of the resources, as possible suppliers. On the other hand, the tone of the provision is very weak in the sense that valuing the rights and interests of the suppliers of genetic resources does not sound as an obligation that actors involved in contract negotiations must fulfil.

Further weaknesses of Decision 391 in respect of the rights and interests of local communities relate to the way the decision provides for the limitations to access and to its enforcement. Indeed, in listing the issues for which the Decision provides for partial or total limitation on access,[37] it fails to include sacred areas and the genetic resources for which local communities can claim to hold significant spiritual values. This is basically a reflection of the idea that, once the legal procedure for access has been complied with successfully by scientific and commercial actors, there is no more concern about the way in which the collection of genetic resources is carried out on the ground. Such collection is legal and can therefore be carried out even on the sacred lands of local communities and will not be violating the law in anyway. The law has also put the power for the pursuit of enforcement operations in the hands of the competent national authority. Local communities do not seem to have been empowered to undertake any judicial process in search of sanctions against any actor whom they may find to have violated the law. Say, for instance, if the communities sign the benefit-sharing annex of an access contract and down the line through the implementation of that contract they realize that the annex is not respected by the scientific or commercial actor, they cannot embark on a legal challenge against this actor themselves. Only the competent national authority can do that, although it is not always obvious that this authority will properly represent and assert the rights and interests of the communities in such processes.

Conclusion

This chapter demonstrates that there are indeed genuine attempts by the international community, regional organizations, indigenous peoples associations and the business community with interests in the assets of local communities to address their concerns within the framework of processes involving community assets. Soft instruments in the form of declaration and professional guidelines have been developed as well as regional regulatory instruments, be they legally or non-legally binding. However, these attempts fall short of accommodating the rights and interests of local communities from a human-rights perspective. Whether they are legally binding or voluntary instruments, they are generally weak in their provisions over such critical rights of local communities as their rights to participation, free and prior informed consent, and rights to undertake judicial proceedings against actors who violate their customary rules and the legally established procedures for access. The regulatory measures provided in the subsequent chapters will attempt to address these weaknesses.

7 Broader framework of the suggested regulatory measures

The foregoing analyses expanded on the limitations of current regulatory regimes such as patents and PBRs systems as well as some national and regional biodiversity regulatory instruments to address the concerns of local and indigenous communities over the scientific and commercial exploitation of their assets. Furthermore, discussions in the previous chapters covered community rights broadly from a human-rights angle and recurrently observed that, community rights have not benefited from the proper consideration that they deserve neither in the biodiversity policy instruments governing access to genetic resources and TK nor in the IP instruments governing exploitation of the outputs of scientific research based on community assets. This chapter presents in general terms the regulatory approaches which, if adopted and implemented by countries rich in bio-cultural assets at the national level, will go a long way to addressing the concerns of rural communities from a human-rights perspective. It does so by proposing the way human-rights principles should penetrate the IP-based biodiversity policy and regulatory tools on access to and exploitation of community assets. The human-rights principles discussed herein are rooted on the international framework of human-rights instruments with specific emphasis on its coverage of local communities and indigenous peoples' rights. The subsequent chapter (Chapter 8) will address in specific contexts the implementation of the policy and regulatory approaches discussed here.

Human-rights dimension of the regulatory measures

Analysis in the previous chapters, especially Chapters 3 and 4 covered discussions on some national and regional biodiversity-related regulatory instruments and regional and international human-rights instruments. Before discussing the rights of local and indigenous communities over their tangible and intangible assets, Chapter 2 dealt with the main drivers of IP systems, especially patents and PBRs systems. When applied in the protection of inventions derived from scientific and technological transformation of the tangible and intangible assets of local and indigenous communities of developing countries, these two systems of legal protection of IPRs have been analysed in respect of their implications to community rights. The short lesson deriving from these discussions is that neither (and understandably so) the IPR systems nor the biodiversity-related regulatory measures accommodate community concerns from a human-rights perspective, that is in the manner reflecting the way community rights and concerns are encapsulated in the relevant international and regional human-rights instruments and approaches discussed in Chapter 4.

One of the aims of this work is to suggest workable IP-based biodiversity regulatory measures, which can be adopted and implemented at the national level, reflecting national realities on biodiversity policy-making and traditional systems of use and management of the resources. The objectives of such regulatory measures must be set by the national policy-making process which should stress conservation of biological resources and preservation of TK and practices of its local and indigenous communities. Furthermore, sustainable use including in respect of scientific and commercially oriented undertakings, which can open up more opportunities for economic development and transfer of technology, need to be one of the main objectives for policy-making at the national level. It is indeed desirable, or rather critical, that the IP-based policy measures reflect a greater understanding of the fundamental principles of IP when, for instance, patents and PBRs are applied in scientific and potential commercial exploitation of community assets. But, in addition, the IP-based biodiversity policy tools must portray respect for the rights of indigenous and local communities as encapsulated in regional and international human-rights treaties. The need to design biodiversity regulatory frameworks and tools to properly address the concerns of indigenous and local communities from a human-rights perspective has already been raised.[1] As discussed earlier, some avenues have also been attempted in a bid to achieve this need. For example, the 1996 submission of Darrell Posey and other contributors to the IUCN Biodiversity Programme, which elaborated on the TRRs concept, contemplates that the protection of the rights of traditional communities in national regulatory frameworks may also aim at achieving the purpose of human-rights protection. The regulatory tools suggested herein should, alongside being seen as practical attempts to craft the protection of the human rights of local and indigenous communities in biodiversity regulatory regimes, also be seen as practical attempts to test the applicability of the TRRs concept that stresses upon human-rights themes such as self-determination.

The critical concerns of local and indigenous communities vis-á-vis their assets that the proposed regulatory tools intend to cover from a human-rights angle include:

- full participation of rural communities at all stages of decision-making processes pertaining to demand for access to and eventual use of the resources and knowledge under their custody and their right to benefit from any exploitation of such assets;
- community empowerment for the determination of how access to and use of their assets should be carried out, especially as the communities require access and use by others to be undertaken in the manner that is respectful of their customary systems of resources management. Such empowerment must enable the communities to veto the projects and activities that they do not agree with;

- community rights of free and prior informed consent and to be consulted over any initiative encompassing access to and utilization of their tangible and intangible assets;
- the proper empowerment of rural communities to have equal possibilities like any other person or legal authority to seek justice and to file legal judicial challenges themselves against anyone who trespasses upon their rights and the established IP-based biodiversity regulations.

All of the human-rights instruments discussed earlier cover these human-rights-related concerns of indigenous and local communities, although regional human-rights undertakings may have varying interpretation and approaches over their implementation. Human principles including equality[2] and non-discrimination,[3] participation in decision-making,[4] rights to benefit from exploitation of the resources under the traditional custody of the communities,[5] self-determination or whatsoever term can be attached to it such as self-identification or local autonomy,[6] are all crafted in the suggested regulatory measures in the manner that reflects and portrays regional interpretations of these principles. From this basis, the promotion of the right of self-determination in these regulatory tools reflects, and is in line with, the African way of interpreting this principle. As commented earlier in Chapter 4 above, African nations can only promote indigenous and local communities rights to self-determination based on the understanding that the enjoyment of this right is based on respect of sovereignty, the inviolability of the borders acquired at the independence of the member states and respect for the territorial integrity of countries.

Below, the IP-based and human-rights-oriented biodiversity regulatory tools are proposed. They are presented in a broad and flexible manner making them adaptable to varying national circumstances. Human-rights principles are tactfully diffused in these tools making them more accommodating and respectful of the concerns of local and indigenous communities. Although these tools are aimed at national implementation, they attempt to cover an option of international enforceability. The practical implementation of these tools in specific national contexts is discussed in the following chapter (Chapter 8), based on selected cases from our study countries: Brazil, India, Peru and South Africa.

Permitting scheme and institutional set-up

The administrative and institutional mechanisms pertaining to the issuance of plant collection permits and research authorizations to scientific and commercial users of the assets of local communities are very critical components of access procedures. These administrative and institutional mechanisms are known as the 'permitting scheme' of an ABS biodiversity policy scheme. Control over and management of these mechanisms is generally concentrated in the hands of a few administrators of forestry and environmental services and research directorates spread across government departments in charge of environment, forestry and research. This approach raises important questions of transparency, governance and participation by other stakeholders such as local communities, whose assets are really the core elements sought by scientific and technologically capacitated actors. Countries rich in bio-cultural diversity and hosting large populations of forest-dependent communities should therefore consider establishing comprehensive permitting schemes that address the concerns and accommodate the expectations of a range of stakeholders having vested interests in PGRs and TK. Stakeholders having vested interests in the biodiversity components are rural communities, scientific and commercial users of PGRs and TK, environmentalists and conservationists, policy-makers and policy analysts. These stakeholders must be integral parts of comprehensive permitting schemes, which must be administered by a trustworthy institutional setup that pulls together a cross-section of capacities and skills needed in the assessment of access permits and research authorizations. The approach of bringing other stakeholders into the permitting processes will go a long way towards addressing the concerns of rural communities in relation to improper access to PGRs and TK; they will address the needs and expectations of scientific and commercial users of these resources by ensuring that access is not cumbersome but can be done in a manner that is respectful to the customary systems and human-rights related concerns of rural communities, such as their right of full participation in the decision-making processes.

An important feature of a comprehensive permitting scheme, which is designed to address the concerns of rural communities and other stakeholders is that the processes leading to the grant of any access permit must undermine irregular access to the assets of local and indigenous communities. Irregular access can be interpreted in various ways. For one thing, irregular access to the assets of local and indigenous communities is access which has been carried out without the legal authorization issued by the relevant issuing authority, the institutional setup referred to above. However, it is my view that in so far as access is done in the way that community rights of participation to the process that led to the grant of the legal permit was not respected, this should be considered as irregular access. In most countries rich in biological and cultural diversity, legal access authorizations in the form of a research permit or a plant collection licence are generally issued to individual researchers, academic research teams, and scientific and corporate users of biodiversity components in order for them to carry out their plant collection and ethno-botanical studies in rural areas without any proper consultation with the community concerned. Ensuring the full participation of rural communities in the decision-making processes pertaining to access authorizations is a key aspect of an improved permitting scheme. Logically, it follows that participation of rural communities in the decision-making processes can be secured through comprehensive consultations that need to involve and be complied with by all stakeholders. Indeed, as long as the traditional societies living in the areas where plant collection and ethno-botanical interviews take place are excluded from such decision-making processes, they will continue to be critical and to feel that their human rights of participation are terribly infringed by both the permitting scheme and the actors involved.

In order to ensure that the above-mentioned irregularities are effectively taken into consideration under an improved permitting scheme, a trustworthy institutional setup should be established. Where necessary and where such an institution is not yet established, and taking into account other important parameters such as the availability of funds and well trained human resources in biodiversity matters, a newly created institution such as a National Biodiversity Authority (NBA) should be established as an independent body. However, a country rich in biological and cultural diversity that has an established institution or a central government directorate already handling permitting matters can forgo the creation of a new and specialized biodiversity authority. The existing service or central government directorate can be transformed into an NBA and enriched with new skills and capacities and thus strengthened to take up heavy and complex duties. The designated biodiversity authority should ensure the coordination of permit-issuing activities and harmonization of permit rules throughout the national territory. However, an important dimension of the institutional setup is the devolution of power of the national authority through the creation or strengthening of the responsibilities of the regional, provincial and local biodiversity offices. At the regional and local level, there needs to be synergies and strong cooperation between the local administrators and representatives of local and indigenous communities, such synergies that will prove very critical in ensuring that the rules are implemented accordingly and that through policy implementation, community rights and concerns are taken into account. Whether or not a new institution has been created or an existing one has just been transformed and strengthened, the existence and functioning of a network of civil society organizations working in the biodiversity sector is crucial in the implementation of the permitting scheme. As watchdogs, the role of the network of civil society organizations is critical at the national level because these organizations can challenge national institutions administering the permitting scheme by bringing their wrongdoing into the spotlight. It is therefore critical that the broader landscape of environmental regulatory policies and broader laws such as those pertaining to good governance in a given country be favourable and supportive to the activities of civil society organizations.

Patenting and benefit sharing

For industries, bio-prospectors and increasingly academic researchers and institutions, patent protection over the products or processes derived from their research based on biological resources and TK is an important aspect of their activities. Patent rights are tools that can facilitate business transactions and can contribute to corporate scientific institutions and commercial entities being able to recoup their huge expenditures in research and development.[7] In this respect, perhaps it is unrealistic for local and indigenous communities and other stakeholders who identify themselves as providers of raw materials to oppose the potential patenting of products or processes derived from the exploitation of biological and genetic resources and TK innovations and practices, during the consultation processes referred to in the previous section. However, in terms of assisting the communities to accept or refuse potential patenting, the purpose of such patenting should be explained to and understood by them during consultations within the framework of the permitting process.

For a country rich in biological and cultural diversity, in addition to its role in market creation, technology development and promotion of creativity, a patent regime should also aim to address the concerns of the communities providing the resources used in research and development, such concerns being related to benefit sharing and sustainable use of the raw materials exploited in R&D processes. In addition, a patent regime should reflect the realities of the context of research activities and innovations at the national and sub-regional levels. Key areas that need to be considered in assessing national and sub-regional realities in research and innovations are the intellectual capacities and skills in terms of trained human resources, the availability of research infrastructure and a functional funding mechanism at the national and regional levels. A patent regime that reflects local research and inventions is more likely to provide greater incentives to national and regional inventors to pursue patenting themselves at the national and regional levels. On the other hand, whether or not patents on biodiversity-derived products or processes are filed at the national or regional patent offices of countries providing the raw materials, and whether or not such patents are filed by and subsequently granted to national, regional or overseas applicants, biodiversity-related patenting should be linked to benefit sharing in one way or another. On this basis, patenting can be seen as a tool that can assist local communities towards the realization of their right to benefit sharing. This is because, in helping corporate bodies create wealth and the products (drugs for instance), scientific and commercial bodies will be in a better position to share the proceeds with the providers of the raw materials. On the other hand, the right to earn proper benefits by local and indigenous communities can be justified on the grounds that they may be able to access newly developed drugs if they are able to afford them or if there are workable licensing mechanisms between developing countries and pharmaceutical companies enabling the former to undertake in-country production of affordable versions of the patented drugs. Furthermore, there should be a mechanism ensuring that patents do not threaten the sustainability of biological resources. Addressing the sustainability of raw materials within the framework of patenting will arguably contribute to communities' fulfilment of their rights to use such resources in their traditional and culturally based systems. In terms of ensuring the sustainability of biological resources, one suggestion is that all stakeholders including rural communities should understand the dynamics that underpin research activities based on PGRs and TK and sustain the integrity of the resources before access is allowed, benefit-sharing agreements are negotiated and patent applications on the products or processes derived from such activities are filed. However, it must be stressed that there is so far no evidence of a direct threat of a biodiversity-related patent to the sustainability of a specific resource which affects the availability of raw materials to traditional communities.

The revision of patent legislations of countries rich in biological and cultural diversity to reflect national realities and address national and regional interests is therefore very necessary. Key areas that should be considered within the framework of such a revision include tailoring the requirements for novelty and inventiveness of patent requirements to reflect the standard of local research and innovations. However, for technology-poor countries with low research capacities, other tools such as the utility models may appear to be most convenient as they are designed for innovations of weak inventiveness. Other important issues that should be considered within the framework of the required revisions are the inclusion of the PIC requirements in the patent procedures. The requirement for the PIC of rural communities can work to ensure that prior to granting a patent on the inventions based on their assets, the registrar at the patent office has evidence that rural communities were informed about the research and potential patent filings and had agreed to these issues beforehand. Where necessary, the 'working' of a patent should also be revised to address national or regional economic and technology development interests. Another important issue to be addressed is the role of biodiversity-related patents in benefit-sharing mechanisms. It is critical that patents or any IP certificate on biodiversity-derived products or processes are profitable to stakeholders in provider countries especially rural communities.

PIC and BIS

In recognition of rural communities' contribution to the conservation and preservation of PGRs and TK, their PIC should be sought before accessing and eventually acquiring and using these elements. Thus, crafting the PIC tool in access processes does reflect the provisions of the CBD[8] and the Bonn Guidelines[9] while promoting respect to community interests. The consent of rural communities must be sought during the consultative process referred to within the framework of permitting procedures. However, the proper informed consent of the communities is needed and is the cornerstone of biodiversity processes because if well implemented, it will improve rural communities' trust vis-à-vis the activities carried out by others on their assets. Throughout the consultative process, all stakeholders should ensure that all segments of the communities are well informed about the aim of the research activities for which their resources are sought. Furthermore, the much-required pursuit of a comprehensive PIC process should mean that there is sufficient disclosure of information by the business partner, the researcher and exploiter of the genetic resources and TK to the providers of these assets. Sufficient disclosure of information is critical in assisting rural communities in making informed decisions about access to their TK and the PGRs on their lands. Local and indigenous communities should therefore be able to know how their assets will be accessed and subsequently used (i.e. for scientific or commercial purposes), whether there will be any application for patent protection over the relevant outcomes, the kind and scope of benefits and compensation they will receive, and whether or not there is a possibility for their assets to be transferred to a third party. The PIC of rural communities should mean that on the basis of the information disclosed by researchers or bio-prospectors, the communities have the rights to accept or refuse the proposed activity. This suggests that if the relevant local and indigenous communities oppose a certain project, the relevant administrative authority should refrain from granting any legal authorization for the related activities. This approach in the realization of the rights to PIC of local and indigenous communities is also aligned with respect to communities' right to self-determination in the way that does not lead to any attempt to undermine neither the territorial integrity of the country nor the power of the authority administering the permitting process. In addition, an important issue that should be taken into account during the PIC process is the fact that the competent authority administering the permit scheme should be required to consult the relevant communities through their elected representatives before making any final decision to grant the legal authorization. Ideally, the consultative process should be based on the balanced information sharing (BIS) approach which suggests that the communities, the users of their assets and the authority issuing the research permit or plant collection licence have the same level of understanding of all activities, are all aware of all operations in as much detail as possible, details and understand and respect each other's concerns.

Enforcement measures

National enforcement: an eye on existing international mechanisms

The enforceability of the policy and regulations on access to and utilization of the assets of local and indigenous communities of countries rich in biological and cultural diversity is often an issue that is addressed somewhat evasively at domestic, regional and even international policy discussions. The way enforcement provisions are tailored in especially domestic biodiversity regulations has led some experts to simply comment that such provisions are overlooked in biodiversity policy and ABS debates.[10] This kind of statement is not absolutely true because the fact of the matter is that national and international biodiversity-oriented regulations have provisions on enforcement. However, such enforcement provisions fail to accommodate acts of infringements of community rights among the punishable offences.[11] On the other hand, the enforcement provisions do fail to empower local and indigenous communities themselves to undertake judicial challenges against any person who might have broken the legally established rules on access and trespassed upon their rights. On these two aspects, domestic biodiversity regulations are not practically effective in respect of their appropriateness to address the concerns of local and indigenous communities.

To be effective and appropriate in addressing the concerns of local and indigenous communities, national enforcement measures that must be included in IP-based biodiversity regulatory measures must recognize that failures by other stakeholders to respect the rights of local and indigenous communities in the context of access processes are punishable offences under the law. These include but are not limited to:

- access by outsiders to undertake plant collection in sacred areas of local and indigenous communities without their consent, and with or without legal authorization;

- failure by the competent authorities to include community representatives in the negotiation and decision-making processes leading to the issuance of the access authorizations (plant collection permits, research authorizations);
- failure by users of community assets to fully disclose intended uses of the resources and the eventuality to seek IP protection over the relevant outputs;
- failure by users of community assets to provide proper compensation to the communities as agreed upon beforehand in the relevant contract.

Not only non-compliance with the law or respect of these basic community rights and concerns should be clearly identified as punishable offences, national biodiversity regulations need to empower the community themselves to seek justice through the national judicial system. Although it is likely that the communities will need technical assistance on judicial matters, they must nonetheless be legally permitted and empowered to initiate court challenges against any actor who trespasses upon their rights. It should be up to the national policy-making process to determine what amounts to offences in relation to disrespect of community rights and to categorize them as civil or criminal offences. It is also up to the national policy-making process to determine the types of sanctions that must be associated with the relevant offences. However, it is critical that sanctions be strong enough to discourage those who might be mistakenly or knowingly trying to play around the rules. For example, the enforcement measures could provide that in the event that someone is caught collecting plant samples in sacred areas, such samples will be taken by the relevant enforcement officers and destroyed and the offender will additionally be fined a considerable amount of money to be paid into a community-development fund. In the event of someone caught repeating such an offence, the offender should in addition to the previous sanctions, be sentenced to an X months to Y years length of imprisonment. Ultimately, punishable offences will not just be limited to illegal access to the PGRs and TK whereby the illegality is only evidenced by lack of a plant collection permit or a legally valid research authorization. The common approach in most developing and biodiversity-rich countries is that competent authorities are generally the sole entities legally empowered to undertake judicial proceedings against offenders. This approach must change to ensure that local communities are legally capacitated to pursue judicial proceedings against offenders, especially when their rights and concerns are at stake in the course of access event if they need help.

The above-described approach to enforcement of biodiversity regulatory measures could be effectively implemented at the national level in respect of cases of infringements and trespass upon community rights by an actor involved in scientific and commercial exploitation of the resources within the geographical boundaries of the country. The big question that follows is how will an overseas actor be prosecuted in the event that community rights are infringed when scientific and commercial exploitation of community assets happens beyond the national boundaries of the source country of the resources? In other words, how should activities that can be described and ascertained as infringing domestic laws of a source country be judicially challenged at the international level and the actors carrying out such activities punished outside the country that provided the raw materials?

In the absence of a legally binding international treaty that reflects national approaches for the protection of the rights and interests of local and indigenous communities, this question proves to be a very difficult one. But, as we attempt to discuss below based on a hypothetical case, it is perhaps not totally impossible to tackle the issue of international enforceability, although even the path we suggest in tackling this issue requires no less than Herculean efforts for any hope of achieving a positive outcome. The difficulty in international enforcement of community rights can be foreseen based on the fact that international environmental treaties are already notoriously known for their weaknesses in dispute settlement provisions. In contrast, the TRIPS agreement is subject to the stringent dispute settlement mechanism (DSM) of the WTO with a definite dimension of trade retaliation for any party that fails to comply with the agreement. With the WTO DSM, one can say that the TRIPS agreement has strong teeth ready to bite any party that does not play by the rules. Unlike TRIPS, international environmental and biodiversity related treaties such as the 1992 CBD and its related and ongoing discussed International Regime on ABS, the 2004 IT-PGRFA and the 1973 Convention on International Trade in Endangered Species (CITES) bear the very classical enforcement and dispute settlement approach. Generally, this approach follows the path of suggesting that parties attempt to resolve disputes amicably in the first instance. After failure to resolve disputes amicably, parties can follow the routes of mediation and eventually arbitration with the treaties generally seeking the goodwill of parties to respect and comply with the arbitral decisions. An important angle that has begun to be mentioned in the dispute-settlement provisions of international biodiversity-related regulatory instruments is the recourse to private international law. This option is among the array of options currently being discussed in the context of ongoing negotiations of the International Regime on ABS under the aegis of the CBD Ad-Hoc Open Ended Working Group on ABS.[12] Both the above-described classical approach to dispute settlement of international biodiversity-oriented treaties plus the additional and gradual inclusion of private international law are unfortunately not backed by strong sanctions in the event of failures of compliance.

There is however a sign of hope in respect of the strengthening of compliance measures of international environmental treaties with the gradual inclusion of options of trade bans in the event of non-compliance by a treaty party. This trend emerges from the recent guide to compliance to CITES provisions adopted at the 14th Conference of the Parties to CITES held in The Hague, Netherlands from 3 to 15 June 2007.[13] It must be stressed that this compliance procedure is crystallised in a non-legally binding instrument which serves a purpose no more as a guide on compliance issues to CITES parties. In addition, Article 2 of this evolving guide invites parties to pay attention on complaince with specific provisions of the Convention, which are identified by Article 2. However, it is worth noting that the convention's provisions on dispute settlement resolution are not among those identified in Article 2. Nonetheless, because the convention's obligations include obligations on parties to abide by its dispute settlement approach, it is our view that this guide applies to Article XVIII of CITES on dispute resolution. Among the measures contemplated by the guide in order to promote compliance, there is a remote, but clearly stated, possibility of suspension of commercial or all trade in specimens of one or more CITES-listed species consistent with the Convention.[14] It will be interesting to monitor how the guide will be implemented in practical terms especially in respect of administering sanctions to parties that fail to abide by the provisions of the convention.

Below we present a hypothetical case of international conflict over access to the assets of developing countries and exploitation of them overseas to underline the difficulties associated with international enforceability of biodiversity regulatory policies. Thereafter, there are analyses of possible avenues through which developing countries and their stakeholders can seek to enforce their rights internationally, using the routes of arbitration and private international law. As stated earlier, going through these routes proves that international enforceability is not totally impossible, although it requires no less than Herculean efforts, which, may, or may not, be worth it.

A hypothetical case of conflicting partnership between a developing Country 'D' and Japan

A research team (Team B) of one university in Japan intends to establish a research partnership with another research team (Team A) based at a university located in a developing country, Country D. This partnership is for the purpose of undertaking research to isolate and characterize some anti-diabetic active molecules from Country D's plants. Team B seeks collaboration with Team A on the basis of the information found in the literature, indicating the involvement of Team A in research on medicinal plants in conjunction with and on the basis of Team A's good relationships with local communities who have been using these plants traditionally in the treatment of many ailments including diabetes in Country D. It must be mentioned that to be able to pursue its research goals, the main source of funding for Team A is through partnering with overseas institutions. Having agreed to the proposed partnership, Team A invites Team B to follow the legal procedures through which a legal authorization for access, collection and research on Country D's biological resources and research partnership is granted.

For the sake of argument, we assume that the permitting scheme of developing Country D's biodiversity regulations are in line with the broader framework of the permitting measures contemplated earlier in this chapter. Considering that these measures suggest full participation of rural communities in the decision-making processes, no access to or acquisition of plant materials should be permitted if the communities have not been consulted beforehand. Furthermore, the materials should only be used in the manner that was disclosed to and agreed by the communities. Any form of utilization of their resources without the consent of rural communities will be considered as an act of infringement to their rights. Also, there should be a signed partnership agreement between all parties before accessing and subsequently using the research materials.

After Team B has complied with the relevant legal procedures and signed a partnership agreement with Team A, the latter collects the required research materials. Preliminary research is carried out in Country D and the outcomes such as the plants' extracts are sent to Team B in Japan for further research. Research carried out in Japan aims at characterizing the molecules and undertaking other necessary tests such as toxicity and biological activity on the extracts initially isolated in Country D. This research also yields other therapeutic properties from these extracts, which were not discovered by or known to Team A in Country D and neither were traditionally known to local communities. Furthermore, Team B is able to demonstrate that the methods applied in Japan are new and bear proven inventiveness and are industrially applicable in the sense of being patentable. With these developments, Team B files a patent application in Japan which is eventually granted over their processes and is looking for industrial partners who might be interested to take up the development phase. The overall process from filing the patent application, through its issuance and the exploration of potential industrial partners, is done without the knowledge of neither Team A nor the communities that provided the plants and TK in the first place. At some stage, an international environmental pressure group, or say a strong civil society organization based in Country D and operating on a global scale, is aware of the commercial exploitation of Country D's plant materials in Japan, investigates and posts useful information on the Internet from where Team A reads it.

In this hypothetical case, we consider Team A and the communities as the plaintiffs and Team B as the defendant. Let us hypothetically assume a lawsuit filed at the High Court of Tokyo by the plaintiffs (likely through a law firm highly reputed for the expertise of its lawyers) stating that the activities carried out by the defendant are essentially commercial, were not discussed initially and therefore breach the partnership agreement and represent an infringement to the rights of the communities that are legally protected under Country D's biodiversity regulations. On that basis, the plaintiffs request the revocation of the patent on the grounds of bio-piracy. However, the plaintiffs may be happy to be properly compensated on the grounds of breach of the agreement, if the patent cannot be revoked on the grounds of bio-piracy. The question is whether any of the requests of the plaintiffs will be heard and given satisfactory ruling by the court in Japan? In other words, how is it possible for the court in Japan to respect the laws of a foreign country such as Country D? Furthermore, assuming that commercial activities may lead to the development of drugs that will be useful in Japan and other parts of the world including Country D, and that there is a ready market for these drugs, is there any grounds for the court in Japan to accept jurisdiction over this case?

IP-related constraints

Some principles of IP law, such as the principle of territoriality, can prove to be very challenging to national initiatives, especially in the absence of an enforceable legally binding international treaty on ABS which promotes respect for the rights of indigenous and local communities protected under international human-rights law. The principle of territoriality of national regulations makes it difficult for a foreign court to address disputes between two parties whereby one party claims that their rights protected under their country's regulations have been infringed overseas. In relation to this hypothetical case, a request to revoke patents granted and exploited in Japan will hardly be given any slim consideration even if the plaintiffs claim that such patents are issued in breach of a contract signed beforehand and may well, pursuant to Country D's biodiversity regulations, be infringing upon the human rights of local and indigenous communities. If it was possible for a foreign court to apply the laws of the country protecting the allegedly infringed rights, this would mean an intrusion of foreign laws into national sovereignty. Thus, in respect of this hypothetical case, the territoriality principle of national regulations makes it difficult for the court in Japan to accept jurisdiction over the case let alone the eventually for making a favourable judgement for the plaintiffs.

Despite the constraint caused by the principle of territoriality, Professor Dinwoodie suggests that it should not be considered as an insurmountable hurdle because the existing international IP regimes provide for exceptions to this principle, for instance, in relation to trademarks.[15] Although the Paris Convention affirms the principle of independence[16] of trademark rights, it provides for the option of extraterritoriality of trademarks.[17] Such exceptions in relation to trademarks rights may be extended to and applied in other contexts, for instance, in terms of ensuring the extraterritorial recognition of biodiversity regulations. However, it appears that the applicability of such an exception in relation to trademarks is highly dependent upon the existence of an international treaty, such as the Paris Convention. Furthermore, the extraterritoriality exceptions on trademarks as provided under the Paris Convention are themselves subject to clauses that allow the state to strongly enforce the territoriality principle.[18] As a result, in relation to our hypothetical case, IP principles represent serious constraints in the plaintiffs' pursuit and hopes for a satisfactory judgement from a Japanese court.

Scenario under private international law

Private international law offers a range of options to the courts in their efforts to deal with international litigations. Broadly speaking, in the case of a conflicting partnership involving partners from different countries, there is a need to decide the court of the country that has jurisdiction over the case on the one hand and to determine the applicable law on the other. In addition, there is a need to ensure that the judgement of the court will be accepted by the counterparts of the other country. Private international law attributes jurisdiction to a national court for a dispute based on a foreign element, determines the applicable law and facilitates the recognition and enforcement of foreign judgements.[19]

The object of private international law is not just to assist and guide the choice of the court/forum and applicable law but is also to ensure that the chosen court/ forum is 'best placed' to hear the case in the interests of justice.[20] In this respect, the selection of the forum for the hearing, which is not necessarily the court, should not affect the substantive outcome of the dispute. Since conflicting cases are generally civil cases[21] by nature, they can be subject to litigation before the court or through arbitration, which is increasingly used as an alternative route for dispute settlement. In order to achieve its object, private international law applies certain factors territorially connected to the case including the domicile of the partners in conflict, the place of registration of the property which is the heart of the conflict and the place of infringement.[22] Establishing jurisdiction over a case is central to private international law and is critical to make justice work for every party. A model known as the 'Savigny model' is acknowledged for its efficiency in the choice of law rule.

The Savigny model seems to have established some degree of codification in the process of establishing jurisdiction over international civil cases. According to Professor Dogauchi, before the Savigny Model the choice of applicable law in international commercial and civil matters was very complex and was worked upon on a case-by-case basis.[23] According to one promoter of this approach, an Italian scholar of private international law, Bartolous Sassoferato (1314–57), the laws relating to personal status should be applied personally. A tentative interpretation of the viewpoint of Bartolous Sassoferato in relation to our hypothetical case suggests that the laws of Country D should apply to its citizens and the laws related to property should be applied in the country providing protection to such a property. Not particularly easy as an applicable approach. Thus, the model of Bartolus did not gain much credit among scholars because it was in reality very unpractical.

The model of the German scholar Friedrich Carl Savigny (1779–1861) gained significant recognition.[24] In his book entitled *System des heutigen romischen* (*System of Modern Roman Law*), vol. VIII, published in 1849,[25] Carl Savigny argues that in establishing jurisdiction over a case, private international law must emphasize the law of the place or country with which the case is most closely connected such as the principal location of business or home country. However, if as a result of the choice of law rules the same law is chosen and applied in every country involved in the conflicting case, then the differences in the content of the laws of these countries do not matter at all. With this approach, the model of Savigny can achieve the objective of private international law in terms of attempting to harmonize cross-border legal differences.

The Model of Savigny for the choice-of-law rule in private international law categorizes conflicting acts and links them with certain connecting factors making sure that acts of the same category are clustered by similar connecting factors. Under this approach, the choice of law rule and the establishment of a jurisdiction over a case seem to be practical. One category of acts may be considered as 'tort'[26] committed in Japan, which for instance may be the use of genetic resources and associated knowledge in a way that is in breach of Country D's biodiversity regulations. It is, however, worth stressing that the existing literature does not provide a clear definition of the concept of 'tort'. This is likely based on the fact that tort can take many forms, including negligence, nuisance, libel, slander, trespass, assault and battery, and it is difficult to provide a definition that accommodates all of these areas. It may therefore be preferable to talk about the 'law of torts', rather than the 'law of tort'. While tort of negligence requires 'negligent' behaviour, other torts such as trespass require 'intentional' or at least 'recklessness' behaviour. 'Torts' can be considered as *wrongs* committed unintentionally or intentionally and some principles such as compensation are associated with tortious acts.[27] The connecting factors are, among others, the place where acts categorized as 'tort' have been committed or the place of nationality of the party that committed such tortious acts or the place where the resources were collected. However, despite the level of codification that the model of Savigny brought in private international law, the choice of applicable law seems to be a matter of national legislation.

The Japanese approach in the establishment of jurisdiction over international civil and commercial matters consists of a three-phase test to be made by the court including:

1. the application of *Jori;*
2. the examination of the jurisdictional grounds in the code of civil procedure (CCP); and
3. the examination of the special circumstances surrounding the case.

For the purpose of establishing that a court has jurisdiction over a case, *Jori* is the norm the judge would create if they were acting as a legislator. On the basis of *Jori,* the judge must assess the case at hand and decide whether or not it can be heard. According to one judgement of the Japanese Supreme Court, *Jori* must reflect the criteria of a fair, just and speedy trial.[28] However, the criteria of *Jori* cannot be fulfilled unless one jurisdictional ground of the CCP Law No. 109 of 26 June 1996 is established.[29] The jurisdictional grounds of the CCP that are commonly taken into account are the place of residence of both the plaintiff and the defendant and the place where the tortious acts were committed. In connection to our hypothetical case, in an attempt to bring about a satisfactory outcome to the plaintiffs, Chapter 8 discusses in a practical manner the extent to which private international law and other alternative routes such as arbitration can work in order to ensure that the rights and interests of the plaintiffs covered by their country's biodiversity regulations are respected overseas. A minimum recognition and respect of the rights and interests of the plaintiffs by stakeholders exploiting their assets overseas will be satisfactory to applaud it as a window of opportunity for the international enforcement of the developing country's IP-based biodiversity regulations.

Conclusion

This chapter presented the broader framework of the IP-based and human-rights oriented biodiversity regulatory tools contemplated in this book. As described, these tools incorporate the rights of rural communities such as their rights of participation in decision-making, benefit sharing, PIC and self-determination, as these rights are recognized and protected by international and regional human-rights instruments. Adapting these broader tools to biodiversity policy-making at the national level in developing countries rich in biological and cultural diversity will go a long way in promoting the human rights of indigenous and local communities in relation to access to and use of their valuable assets. In addition, the adaptation of these tools to national-level policy-making will maintain due respect to other forms of rights such as the IPRs of scientific and corporate users of community assets. It is my view that the promotion of community rights in biodiversity regulatory measures and their respect by scientific and commercial users of community assets will improve the trust of local and indigenous communities in the activities of scientific and corporate bodies and could incentivize them to control access and promote conservation and the sustainability of their valuable assets in the interests of humanity.

8 Applicability of the regulatory measures

The regulatory measures for access to the assets of indigenous and local communities in developing countries rich in biological resources and the associated knowledge, innovations and practices of local communities described in the previous chapter are very broad in nature. They are aimed at all developing countries, not just the four covered in this book, and must be adapted to each country's own environmental, social and economic circumstances. The thrust of these measures is that they are framed to assist stakeholders involved in all access processes to ensure that respect of the rights of local and indigenous communities is embedded in their actions and that community needs and expectations are central in IP-based biodiversity policy-making. This chapter discusses how some of these measures can work in practice in the study countries in order to achieve the aim of protecting community rights and addressing their concerns from a human-rights perspective. Furthermore, despite its recognition of the constraints associated with monitoring compliance with these measures when community assets are scientifically and commercially transformed beyond the geographical boundaries of the providing developing countries, discussions in this chapter attempt to cover the possibility of a wider applicability of the proposed measures (beyond the four study countries) and their international enforceability.

Permitting scheme and institutional set-up: community participation in South Africa

The principle of participation in decision-making processes by indigenous and local communities has been discussed in this work as an important human-rights principle which must be recognized and protected. Participation is not just a critical human right for local communities, it is also key to the pursuit and realization of the principle of sustainable development. So, while developing policy measures and practical mechanisms aimed at securing full participation of local and indigenous communities in the decision-making processes pertaining to the issuance of the relevant permits for access to their assets, stakeholders should bear in mind that their actions are also contributing to the realization of sustainable development. The principle of participation has gained international acceptance as a sustainable development element as evidenced by its inclusion in the works of the UN agencies. The UN FAO endorsed the 'Peasant's Charter' in 1979, which provided that: 'participation by the people in the institutions and systems which govern their lives is a basic human right, and also essential for realignment of political power in favour of disadvantaged groups and for social and economic development'.[1] Just over a decade later, the 1992 Rio Earth Summit adopted the Declaration on Environment and Development which included a key principle on the issue of participation. Principle 10 of the Rio Summit provides that: 'environmental issues are best handled with the participation of all concerned citizens at the national level'.[2]

While stressing the specific role of indigenous peoples in the realization of sustainable development, Principle 22 provides that: 'indigenous people and their communities and other local communities have a vital role in the environmental management and development because of their knowledge and traditional practices. States should recognize and duly support their identity, culture and interests and enable their effective participation in the achievement of sustainable development.'[3]

As discussed in the previous chapter, the current policy and legal framework regulating access to community assets in South Africa is failing to provide for local communities' full participation in the decision-making processes within the framework of access to their assets by others. We also discussed approaches concerning the establishment and proper empowerment of an institution responsible for coordinating and overseeing all the tasks and consultations related to the procedures through which legal research authorizations and plant-collection permits are granted. An important dimension of the approach we suggest is the manner in which it contemplates and secures indigenous and local communities' full participation in the processes as key stakeholders and on human-rights grounds. In 2004, South Africa enacted a new law, the Biodiversity Act No. 10, to provide for the management and conservation of South Africa's biodiversity in line with the National Environmental Management Act 1998. The Biodiversity Act No. 10 2004 has a broad coverage, which extends to:

- the protection of species and ecosystems that warrant national protection;
- the sustainable use of indigenous biological resources;
- the fair and equitable sharing of benefits arising from bio-prospecting involving indigenous biological resources; and
- the establishment and functions of SANBI as the biodiversity authority of South Africa pursuant to Chapter 2 of the Biodiversity Act No. 10.

In particular, Chapters 6 and 7 of the Act deal with bio-prospecting activities and the use of South African indigenous peoples' knowledge associated with the country's biological and genetic materials in research that might or might not have commercial ends. These chapters lay down the rules pertaining to the permitting procedures. These procedures bear significant strings in respect of the application and potentially the acquisition of a research permit; they outline the criteria for a permittable biodiversity-related research activity that targets the assets of indigenous communities in South Africa and identify the authorities that have the power to grant the relevant permits in South Africa. However, the permitting rules established by the Biodiversity Act No. 10 of 2004 were not in force until April 2008, as they awaited the development and enactment of the Biodiversity Regulations[4] which were completed in February 2008 and came into force in April 2008. Thus, while paving the way for the implementation of the permitting rules pertaining to bio-prospecting and ABS activities pursuant to Chapters 6 and 7 of the Biodiversity Act No. 10 2004, the coming into force of the Biodiversity Regulations of 2008 theoretically ended the patchy approach in regulating bio-prospecting activities which were administered across the country by the provincial conservation boards.[5] Against this background, let us now discuss the extent to which existing institutional and regulatory frameworks on ABS in South Africa take into account local and indigenous communities' rights of full participation in the relevant permit procedures. Alongside this assessment, we provide some tentative avenues, the implementation of which will go a long way in ensuring community participation, thus materializing their human right to full participation to decision-making processes on issues affecting them.

A critical assessment of the permit system currently in force in South Africa reveals that it does not adequately ensure the participation of rural communities in the decision-making processes through which permits are approved and granted by the designated issuing bodies, notably the minister in charge of environmental issues for permits related to bio-prospecting research and the MEC for permits related to research other than bio-prospecting.[6] Indeed the Biodiversity Act No. 10 of 2004 and the implementing regulations of its bio-prospecting and ABS rules, the 2008 Biodiversity Regulations, fail to ensure that traditional societies living in areas where plant collection and ethno-botanical interviews take place participate in the decision-making processes through which a legal authorization for research or a plant-collection permit is issued. In this respect, the extent to which the issuing authority will be able to ensure that researchers or bio-prospectors, to whom the permits are granted, will respect the local systems of resources management of traditional societies and address their expectations during their activities is highly questionable. Let us justify this rather succinctly. The mechanism contemplated in Parts 2 and 3, Regulations 9–13 of the 2008 Biodiversity Regulations have concentrated the power of the administration and handling of the permitting scheme in the hands of highly centralized issuing authorities. All applications must be submitted to these authorities and the supporting documents, including and where necessary, proof of benefit-sharing agreements, proof of material transfer agreements and proof of identification, and consultations with other stakeholders such as the affected communities must also be submitted by the applicants before (or as a condition for validity) the permit is issued by the issuing authority. However, there is no provision in the regulations to address how permit applicants should go about signing the necessary agreements and consulting with local communities, neither is there any mechanism crafted in the 2008 Regulations that can aid the issuing authority to check that any supporting document or proof was lawfully acquired. The issuing authority will very simply use the proof as provided by the applicant, and, if such proof is satisfactory, the decision about the grant of the required permit is likely to be positive. The 2008 Regulations have no provisions on the devolution of some power to bodies at the provincial or local level, which, if entrusted with some amount of responsibility within the framework of the permit procedures, will assist in monitoring compliance by permit applicants to such critical aspects of the permitting processes as the conduct of proper consultations with local communities.

Thus, as hinted above, while harmonizing the permitting rules across the country, the Biodiversity Act No. 10 of 2004 and its 2008 implementing Biodiversity Regulations are failing to achieve proper respect and protection of community right to full participation in ABS processes. In addition to achieving harmonization of the permitting rules, the setting up of an institution to administer the permitting scheme at the national level is another step which can aid the fulfilment of the required changes in respect of community participation. Section 10 of the Biodiversity Act No. 10 of 2004 provides for the establishment of a national competent authority, defined as the National Biodiversity Institute. To achieve efficiency in the handling of the administration of the permitting rules, practically speaking, the National Biodiversity Institute should handle the tasks and responsibilities related to permitting procedures. Ideally, the National Biodiversity Institute should establish provincial representations, entrusting them with the responsibilities to receive and examine applications for access authorizations, approving and issuing permits for research other than bio-prospecting research (scientific and academic) with the consent of the communities, if the proposed activities are likely to have some impacts on identifiable communities. However, for permit applications related to bio-prospecting activities or scientific research with potential commercial applications, the provincial representation of the National Biodiversity Institute should be expected to forward the application to the national authority that will make the final decision for grant or refusal of the research authorization or plant-collection permit. In any case, whether or not a permit is issued at the provincial level, copies of applications and appropriate notifications should be forwarded to the national authority. In order to handle such responsibilities both at the national and provincial level, the National Biodiversity Institute must have access to a pool of skilled human resources. The required skills include strong botanical assessment capacities, strong negotiation skills, sound understanding of the commercialization of products resulting from biodiversity research, and sound understanding of the implications of IPRs in ABS processes and the scientific and commercial use of community assets.

There are obvious difficulties in relation to assigning the above-mentioned responsibilities to the National Biodiversity Institute. For one thing, the Biodiversity Act No. 10 of 2004 has not assigned such responsibilities to the National Biodiversity Institute. The Act establishes the National Biodiversity Institute as an advisory body, to provide recommendations to the minister in charge of environmental issues regarding application for access authorizations. The Biodiversity Act No. 10 of 2004 gives a great deal of decision-making power to the minister, as reflected in the long list of regulations he or she is entitled to decide upon; for example, 'the designation of State's organs as permit issuing authorities'.[7] Consequently, not only did the Biodiversity Act No. 10 of 2004 narrow the decision-making power of the national authority, it also failed to identify the central administration that would be responsible for the permit system. The minister in charge of environmental issues is even empowered to discretionally make certain decisions, meaning that, in using such discretional power in respect of issuing a bio-prospecting permit or approving a benefit-sharing agreement, it is highly likely that other stakeholders including local communities will have no to very little say in the process. Furthermore, the SANBI does not even appear to be the appropriate institution that should handle the responsibilities of the permit system because this institution cannot at the same time act as judge and party. Indeed, the SANBI is largely known for its involvement in bio-prospecting activities in South Africa.[8] Also, the SANBI does not have the required skills mentioned above. Moreover, as highlighted by one policy expert who has been observing the evolvement of biodiversity policy in South Africa, 'SANBI may not even want to or accept to handle such responsibilities'.[9]

In light of these difficulties, a second suggestion is to identify a coordinating body, which could be an existing directorate (at the Department of Environmental Affairs and Tourism (DEAT)) and assign it with the responsibilities associated with the administration of the permitting scheme. The existence of strong civil society organizations in South Africa will ensure that the activities of the identified authority at the DEAT are consistently monitored and properly criticized. Well-known NGOs established and operating in South Africa that can perform the monitoring role of the activities of decision-making bodies on environmental issues in South Africa include Biowatch South Africa,[10] and the African Institute of Corporate Citizenship, which is an NGO committed to promoting responsible growth and competitiveness in Africa by changing the ways that companies do business in Africa.[11] If the overall administration of the permitting scheme is entrusted to a directorate of the DEAT, it could be more practical to leave the implementation of the permitting rules and procedures with the provincial conservation departments. However, there should be a division of responsibilities between the central and provincial services. The central authority at the DEAT will only be responsible for issuing bio-prospecting-related permits while the provincial nature conservation departments will be issuing non-bio-prospecting research permits, according to the regulatory framework in force. Considering the need to ensure full participation of rural communities in the decision-making processes, how is the model going to work in practice?

Applicants will submit their permit applications for bio-prospecting and non-bio-prospecting research activities to the provincial nature conservation departments. All applications will be assessed at the provincial level in order to establish whether they are aimed at pure scientific or academic research activities, bio-prospecting, or scientific research with potential commercial implications, in line with the definitions of bio-prospecting research and research other than bio-prospecting provided by the 2008 Biodiversity Regulations. Permit applications must clearly indicate the types and quantities of resources needed, the areas where the materials will be collected and whether the knowledge of the communities will be sought through interviews during field work. If it is established that the permit application is for research other than bio-prospecting, it will be wholly processed and possibly approved by the provincial nature conservation department. In this case, permit applicants will be urged to conduct consultations with the specific communities of the areas they intend to undertake field work. In the course of such consultations, they will assure the communities that their knowledge and resources are sought for research only and that they will be acknowledged in the final reports or in the resulting journal articles as the providers of the raw materials. This is an important approach because it will give permit applicants the opportunity to provide the communities with clarifications about their real intent and will give the communities the opportunity to understand the issues at stake and decide whether to accept or refuse the research activity if they find it disrespectful to their traditional systems of resources use and management. The report of such a consultative process, clearly establishing the consent of the communities, will be sent to the provincial nature conservation department and will form the basis for the final approval or refusal of the permit. If the project or activities for which the permit is required are academic research only, there need not be any requirement to identify all the communities that might be sharing the same knowledge because this would be burdensome to academic research. Equally, there need not be any requirement for a comprehensive benefit-sharing agreement because the activities have no commercial purpose. Likewise, there should not be any requirement for an environmental impact assessment (EIA) because the required materials are likely to be collected in very small quantities; the permit applicant can also be exempted from the requirement to invest in ex-situ or in-situ plant conservation. However, the permit will only be issued when the applicant has paid the required fees, R100 as stipulated in Annexure 1 of the 2008 Biodiversity Regulations. Whether or not the permit is granted, the provincial nature conservation department will send copies of applications to the central authority at the DEAT where a database will be developed and consistently managed, allowing proper monitoring of biodiversity research activities throughout the national territory.

If it is established that the permit application is for bio-prospecting activities or that research is likely to lead to some forms of commercial applications, for instance, research conducted in the academic context leading to the production of cosmetic or other unpatented phytomedical products, preliminary assessment of the application will be carried out at the provincial level but the central authority at the DEAT must be involved in order to ensure that certain issues are addressed before the permit is eventually granted. The central service will ensure that applicants for bio-prospecting activities have developed a comprehensive benefit-sharing agreement with the communities whose knowledge and resources are targeted. The communities living in the specific area where field activities are carried out will be primarily consulted. However, in order to ensure that potential conflicts among communities do not arise in the future, all the communities likely to share the knowledge must be identified. The first communities consulted will be used to identify other communities while the involvement of social scientists such as anthropologists will also be useful. As plant collection for commercial use can be detrimental to the sustainability of the resources, the EIA should be carried out by the SANBI. The central authority at the DEAT will ensure that the bio-prospecting actor has committed to invest in in-situ and ex-situ conservation of the resources sought. Overall, all of such processes will be carried out at the expenses of the permit applicant.

During consultations with the communities, permit applicants will clearly present their project and disclose the prospective commercial activities, including any intent to file patents. The consultative processes will give the opportunity to the communities to accept or refuse the proposed activities. Staff of the central authority at the DEAT or the provincial nature conservation department will facilitate these consultations and will assist the communities to understand the issues at stake and to make informed decisions. Once the central authority is satisfied with the outcomes of all these consultations based on the proof provided by permit applicants on the one hand and on information gathered through its interaction with the provincial bodies and local communities (through their representatives), it will either issue the permit or request the provincial department to do so. This approach will ensure that permits are granted only if the communities have agreed upon the activities that will be using their assets and that such permits are only granted when all the conditions are met in a proper manner and that permits are not granted unconditionally[12] by the issuing authorities.

Obviously, one can argue that this system is bringing additional bureaucratic burdens that will likely slow down research activities in South Africa. Such arguments have been put forward in other instances such as in the case of Botswana, where it is argued that the current administrative procedures regulating plant collection for research purposes are lengthy and complicated and have slowed down academic research.[13] It is worth underlining that the process of filing and eventually being granted a patent for an innovation based on biological resources is complex and very expensive. Patenting is such an important dimension of R&D organization that a patent portfolio for instance plays an important role in the worth of the organization. For commercial intents, business and partnership development purposes, R&D actors are ready to go to any length to secure patents or even PBRs on the outputs of their research undertakings based on community assets, despite all the constraints and expense of the process. Community assets hold important cultural, social, economic and spiritual values to local communities. And they are increasingly critical economic assets to countries hosting them and the communities. There is no intention whatsoever for the permit scheme discussed here to complicate access to PGRs and TK in South Africa or in any developing country. Instead the issue here is essentially about protecting and respecting community rights to full participation and empowering them to as much as possible control access to their assets because they are in a better position to promote biodiversity conservation. Strictly abiding by the process contemplated here will increase transparency and trust between local communities, policy-makers and potential users of community assets in R&D undertakings. The ultimate goal of the process is to ensure that community assets are used for the good of humanity in respect of the rights and expectations of local communities who are ready to share. Any actor who complies with the process will find it short. But trying to play tricks with the process will obviously lengthen it and will potentially taint the credibility of any actor attempting to do so.

Patents, community concerns and benefit sharing in South Africa and Brazil

South Africa is now well integrated into the international community. One evidence of this global integration is the country's commitment to make its national laws reflect international treaties such as the WTO-TRIPS Agreement, the 1992 CBD, the International Convention for the Protection of New Varieties of Plants 1978 and 1991 and the FAO IT-PGRFA. The South African Patent Act No. 57 of 1978 has provisions on the patentability of products or processes resulting from bio-technological activities. Section 25(1) of the Patent Act No. 57 of 1978 provides that, for a product or process deriving from biotechnology activities to be patentable, it must be new, result from an inventive step and be industrially useful. Regarding what amounts to unpatentable subject matters, Section 25 of the Patent Act 57 of 1978 provides that they include a discovery, scientific theory, mathematical method, literary, dramatic, or musical innovation, program for a computer, invention that the publication or exploitation of which would encourage offensive or immoral behaviour, any variety of animal or plant, or any essentially biological process for the production of animals or plants not being a micro-biological process or the product of such a process.[14] In relation to the patentability of plants or animals, the Patent Act No. 57 of 1978 provides that these elements are patentable subject matters provided they derive from micro-biological processes. For instance, transgenic plants and animals are patentable if they meet the three fundamental requirements for patentability.[15]

'Novelty' and 'inventive step' are two very important requirements for the patentability of any given invention, and they are assessed against the 'prior art' that is related to a claimed invention. As it appears to be the approach of the South African patent system, the assessment of the prior art seems to be based on information and data attuned to modern science, thus excluding TK and practices of rural communities from the prior art concept and making it hard to potentially challenge a biotechnology patent.[16] Section 25(10) of the 1978 Patent Act No. 57 of South Africa, substituted by Section 31(d) of the Intellectual Property Laws Amendment Act No. 38 of 1997, contends that the inventiveness of an invention is linked to any matter forming the state of the art. The state of the art of an invention is composed of the common knowledge regarding the alleged invention immediately before the priority date (for patent law purposes, the date which is assumed to be the date of invention). According to J. Younger, in *British Thomson-Houston Co. Ltd* v. *Stonebridge Electrical Co Ltd.* (1916), common knowledge is the information which, at the filing date of the patent in question, is common knowledge in the art of science to which the alleged invention relates so as to be known to duly qualified persons engaged in that art or science.[17]

This approach in the assessment of prior art for the purpose of establishing inventiveness practically disqualifies the oral and even makes it difficult for written TK, innovations and practices of local communities to defeat the inventive step criteria of a scientifically proven and biotechnology-based invention.

Let us now focus on the assessment of 'novelty'. Section 31 of the Intellectual Property Amendment Act No. 38 of 1997 substituting Section 25 of the Patent Act No. 57 of 1978 provides that 'an invention shall be deemed new if it does not form part of the *state of the art* immediately before the priority date of any claim related to that invention'. Elements required in the assessment of the *state of the art* as per Section 25(6) of the Patent Act No. 57 of 1978 shall comprise all matters (whether a product or process or information about either, or anything else) which has been made available to the public (whether in the republic or elsewhere) by written or oral description, by use or in any other way. Theoretically, this provision suggests that with regards to biotechnology-based inventions, prior use of a plant and the prior knowledge held by rural communities could be considered 'novelty destroying occurrences', because they can be considered as part of the state of the art. But practically speaking, the general principles applied by patent examiners in the assessment of novelty suggest that novelty can be defeated, thus invalidating the patent if the standards relied upon are substantially identifiable to the invention. However, if there is a difference, which is more than practically immaterial, albeit not inventive, then there is novelty.[18] The extent to which TK and practices of rural communities may be substantially identifiable to a biotechnology innovation, thus defeating a related patent, will be hard to demonstrate.

The standards applied in the assessment of the novelty of the invention otherwise known as 'novelty destroying occurrences' include prior knowledge, prior publication or prior use, all of which are based on *reasonable technical trial or experiment* undertaken by the applicant or any person or persons from or through whom the applicant has derived their rights. By reasonable technical trial or experiments, it is implied that, in order to defeat the novelty of an alleged invention, they should be modern science-related. In this respect once bio-resources and the associated TK have been acquired by researchers and transformed in some way through modern science and technology, it does not fall anymore into the realm of the state of the art, because such technological knowledge was not known to and/or used by rural communities. As a result, 'prior use' and 'knowledge' of rural communities related to the raw biological materials will hardly be able to destroy the novelty of a scientifically proven and technology-based invention.

An interesting feature of the South African patent system is its approaches in addressing the 'working of a patent' and 'compulsory license'. Although the patent system of South Africa does not exclude the importation of foreign patented products into the South African market, the working of a patent in South Africa seems to stress the technical and commercial exploitation of the said patent in South Africa. Article 56(1) of the Patent Act No. 57 of 1978 provides that a compulsory licence can be issued to an applicant if there are evidences of abuse of patent rights. Articles 56(2)(a)–(e) of the Patent Act No. 57 of 1978 clarifies what constitute the elements of abuse, which include non-working of a patent in the Republic of South Africa, importation that may hinder working in the Republic of South Africa, demand of patented products not being met, prejudice to local interests, and importation with excessive price. Section 45 of the Intellectual Property Laws Amendment Act No. 38 of 1997 amending Section 56 of the Patent Act No. 57 of 1978 provides that, in considering the grant of a compulsory licence, the commissioner would assess the application on its merits and may order the grant of a licence on conditions that can prevent the applicant from importing the products into South Africa. Altogether, these provisions suggest that importation of patented products is not an important aspect of the working of patent in South Africa while local working of a patent is.

However, despite its interesting approach in addressing the working of a patent, the patent system of South Africa does not reflect the necessity to conserve and preserve the rich biodiversity of the country as it fails to address the concerns of rural communities. In the present context, there is no indication that, prior to filing a patent application, the applicant must seek the consent of rural communities if his or her invention is based on community resources and associated TK practices. However, a significant step was made with the enactment of the Patent Amendment Act No. 20 of 2005. The purpose of this Act is to amend specific sections of the Patent Act No. 58 of 1978 so as to insert certain definitions and to require an applicant for a patent to furnish information relating to the contribution of an indigenous biological resource, a genetic resource or TK or practices in an inventive process and to provide for matters therewith.[19] Section 3(A) of Act No. 20 of 2005 requires every applicant for a patent to lodge before the registrar, and before the patent is granted, a statement in a prescribed manner stating whether or not the invention for which protection is sought is based on or derives from an indigenous biological resource, genetic resource or TK or use. The submission of such a statement is, however, not enough to satisfy the registrar in respect of the possibility of granting the patent, because, according to the Act, upon receipt of this statement, the registrar is allowed to request further proof from the applicant in support of his or her statement.[20] This is where, arguably, proof of the various disclosures should be provided, including proof that the applicant's authorization to use the indigenous biological resource and TK and to eventually patent its derivatives derives from prior disclosure of such intents to and the acquisition of the consent of local and indigenous communities. If there is a failure to provide these proofs, the registrar should be able to refuse or delay granting the patent. It must however be stressed that the provisions of the Patent Amendment Act No. 20 of 2005 are not part of the substantive patent criteria in South Africa. Thus, while failure by an invention to comply with one of the three main criteria for patentability means automatic failure to the patentability test and no grant of patent protection over the claimed invention, the provisions of Act No. 20 of 2005 may not be consistently monitored by the patent registrar. It will thus not be surprising if many patents are granted in disregard of the various proofs that need to be provided as required under Section 3(B) of the Patent Amendment Act No. 20 of 2005.

The Patent Amendment Act No. 20 of 2005 is in conformity with the broader indigenous knowledge system (IKS) policy that was adopted in 2004 in South Africa. This policy was the result of an interdepartmental effort to set up a tool for the recognition, understanding, integration and promotion of South Africa's wealth of indigenous knowledge resources. The IKS policy contemplated the protection of indigenous knowledge and the holders of such knowledge against exploitation, and recommended the development of legal and administrative tools that could aid the protection of indigenous knowledge and its holders. In a drive to put the recommendations of the IKS policy into practice, the Department for Trade and Industry (DTI) produced a report in 2008 on the protection of indigenous knowledge using the IP system. This report discusses opportunities, benefits and shortcomings in respect of the application of the various IP tools for the protection of indigenous knowledge resources. In relation to patents, the DTI report contends that for the purpose of protecting indigenous knowledge resources and its holders, if an invention occurred because it was based on local communities' assets, the various disclosures shall be complied with by the patent applicant. In addition, the report requires that there must be benefit-sharing agreements in place as well as the possibility of co-ownership of patents.

Like South Africa, Brazil has a very well established patent system, which reflects the major international treaties on intellectual and industrial property. In fact in enacting its very first patent law as far back as in 1809, Brazil was the fourth country in the world to establish a patent regime. Brazil was one of the founding members of the 1882 Paris Treaty on Industrial property and has remained a very active member of the treaty since. Like South Africa, Brazil is very active in international forums where patent issues are discussed, both broader and substantive patent issues as well as on topical issues such as patents in respect of scientific and commercial exploitation of biological resources and TK of local and indigenous communities from developing and biodiversity rich countries.

The most recent IP law of Brazil is the Industrial Property Law 14/05/1996 No. 9.279. This law covers the various IP tools including patents and utility models; industrial design; trade marks and geographical indications. Title I of this law addresses patents. Without engaging in an extensive analysis of the 1996 Brazilian law or its chapter on patents, our brief commentary focuses on the strengths and weaknesses of the patent provisions of the Brazilian law in addressing the concerns of rural communities over issues such as 'bio-piracy' and 'benefit sharing'. Like many other patent regimes around the globe and very similar to the South African approach discussed earlier, Title I of the Brazilian industrial property law makes provisions on the criteria for patentability, defines what is and what is not patentable and addresses the concept of working patents. Thus, like many patent regimes, an invention is patentable in Brazil if it is new, derives from an inventive step and can be exploited industrially.[21] Non-patentable inventions are outlined in Article 10 of the law and include, among other items:

- discoveries, scientific theories and mathematical methods;
- purely abstract conceptions;

- commercial, accounting, financial, educational, advertising, raffling, and inspection schemes, plans, principles and methods;
- literary, architectural, artistic and scientific works, or any aesthetic creations;
- computer programs;
- presentation of information;
- rules of games;
- surgical techniques and methods, as well as therapeutic or diagnostic methods, for application to human or animal bodies;
- all or part of natural living beings or biological materials found in nature, even if isolated there from, including the genome or germplasm of any living being and the natural biological processes.

In defining what kind of invention can be considered as new for the purposes of patentability, Article 11 of the Brazilian law provides that an invention is 'new' if it is not part of the state of the art. Further, Article 11(1) describes the state of the art as consisting of everything that became accessible to the public prior to the filing date of the patent application, by means of a written or oral description, by use or by any other means, in Brazil or abroad. A further elaboration on the state of the art for the purposes of determining the novelty of an invention is provided under Article 11(2) which stipulates that, for the purposes of determining novelty, the entire content of an application filed in Brazil and not yet published shall be considered to be the state of the art from the date of filing or of claimed priority date, provided that it becomes to be published even subsequently. Another important provision of the Brazilian law is Article 45 introducing the right of prior user of an invention. According to this provision, if a person was already, before the filing date or the date claimed as priority date, and in good faith, using or exploiting the objects over which a patent is filed, such person shall continue using the relevant objects even if they are now protected by patents.

To a certain extent, and at least theoretically, it may be argued that the Brazilian patent regime is made of strings that can be used to address the concerns of local and indigenous communities in respect of patenting over their assets. For one thing, the Brazilian approach in describing the state-of-the-art concept sounds very broad and could well stop patent applications over inventions based on biological resources and TK which are very similar to the way indigenous and local communities use their assets. This is because the state of the art consists of written and oral materials. Furthermore, Article 10 of the law very specifically excludes from patenting mere discoveries, biological materials or their parts, and parts isolated from biological materials. Also, local communities can well be protected by provisions on prior user as this provision means that they theoretically have the rights to continue using their biological materials and TK that might have formed the basis of a patented invention. It is, however, not clear if this means that local communities can continue commercializing their products if they were doing so before the filing date of the patented invention and even if such continuation means significant losses for the patent holder.

Despite these strengths, it does not seem that based on the law, patent proceedings in Brazil are shaped in a way that really addresses the concerns of local and indigenous communities. In essence, like the South African patent regime, the Brazilian law fails to accommodate the concerns of local and indigenous communities, essentially making patents work for technologically capacitated actors with less regard to local and indigenous communities. Section III of Title I of the law deals with processing and examination of the patent application. Article 35 of the law stresses upon 'technical examination', stipulating that a report of a search and an opinion shall be prepared in respect of matters such as the patentability of the application; appropriateness of the application given the nature of the claimed invention; reformulation or division of the application; and technical requirements. It appears that the report resulting from the search and examination process shall be of a technical nature, which will play a critical role in the determination of the patentability of the claimed invention. It is therefore difficult to see how oral information that amounts to prior art that could be provided by local communities in a bid to stop what they would see as a bio-piracy patent would be able to defeat the novelty of newly proven technical processes claimed in an invention if such processes are deemed to have complied with the novelty requirement of the patentability criteria. On the other hand, the proceedings contemplated in the context of search and examination of a patent application do not even identify local and indigenous communities as parties who may contact the examiners in order to submit information or data that may help the assessment of the prior art. Arguably, local communities can use the provisions on judicial nullity proceedings to seize the court in demanding that a given patent be nullified if they do not find it to be compliant with the novelty criteria for instance. But, as is always the case, judicial proceedings can be very costly and beyond the affordability of many local communities. They may not have the muscle to exploit judicial proceedings to stop bio-piracy patents.

One issue that is very surprising, especially in the case of Brazil, is the fact that given the very active role of this country in advocating the inclusion of the requirements for disclosures in patent applications in the meetings of the TRIPS Council, the 1996 Intellectual Property Law is very silent on this issue. Unlike South Africa, this country has not yet enacted a law on the disclosure requirements in patent applications. This means that Brazil has not yet taken a legislative approach to address such human-rights-related concerns of rural communities as benefit sharing within the framework of patents. Our suggestions below concerning the way that developing countries can reformulate their patent approaches to better accommodate the concerns of local communities are aimed at both Brazil and South Africa and are obviously adaptable to the circumstances of many developing countries.

Patent systems of South Africa and Brazil are well established and are effectively being used by research institutions and industries in these countries and across the world. On these grounds, the patent systems of these countries are being employed in the valorization of research and innovative efforts at the national level, making these countries attractive to the international scientific community and important partners in the global trade. However, knowing that Brazil and South Africa are megadiverse countries, the richest territories in biological and cultural diversity and that components of their biocultural diversity richness are being exploited for commercial purposes, it is surprising that their national biodiversity laws[22] attach very little to no attention to IP, especially patents. To address the concerns of local and indigenous communities in Brazil and South Africa in connection to issues such as benefit sharing and bio-piracy, it is important to carve out mechanisms directly through their patent regimes.

In terms of addressing the concerns of rural communities within the framework of patents, it is important that the communities understand the purposes and key principles of patents, especially in respect of the protection of products or processes based on the exploitation of their assets. The consultation processes discussed in the previous section within the framework of the permitting system offer a good avenue for such discussions, during which the communities must be able to accept or refuse patent application on the products or processes derived from their assets if such patenting is harmful to their cultural integrity. The requirement of a patent applicant to submit a statement and proofs outlined in the Patent Amendment Act No. 20 of 2005 of South Africa reflect this approach and must aim at empowering the registrar to ensure that in submitting the proofs of disclosures, the patent applicant has respected the views of the relevant communities. The disclosure of origin should not just stand as a simple provision requiring patent applicants to provide information about the origin of the genetic resources and associated knowledge used in the research processes, but should also, and importantly, require them to provide evidence that the communities were properly informed about the research activities in question including the possibility of filing a patent application over the resulting products or processes. These evidences must be provided in the report of the consultation processes during which the PIC of rural communities is sought.

The assessment of 'novelty' and 'inventiveness' of an alleged invention based on PGRs and TK, as provided in the patent regimes of Brazil and South Africa, is heavily based on scientific and technical information associated with the claimed invention. Therefore, like the patent systems in the developed countries, the Brazilian and South African legislations seem to facilitate the privatization of TK and biological resources. However, perhaps such privatization is understandable because patent systems operate just the way that they are supposed to. Nonetheless, in terms of addressing the concerns of rural communities under the patent system, the assessment of 'prior art' or 'state of the art' in relation to 'novelty' and 'inventive step' of an alleged invention should take into account TK, innovations and practices of rural communities, whether published or not. The novelty of the invention and its inventiveness should be destroyed, thus defeating the patent if, during the examination process, the registrar at the patent office finds that patent claims are highly similar to the prior knowledge and practices of rural communities. One problem with this suggestion is that it might be nailing the feet of rural communities themselves, because, if it was possible for them to seek patent protection over their own technologies, these rules would stop them from achieving such protection. But the fact of the matter is that TK and technologies associated with PGRs and held by rural communities are not patentable. In this respect, if implemented, this approach in assessing the 'prior art' concept of a patent application may not be detrimental to rural communities, but may perhaps have an affect on 'bio-piracy'.

When patented, the products or processes derived from scientific research based on PGRs and associated TK are likely to be of interest to commercial operators for the purpose of development into consumer products. Such products may be addressing the needs of the larger public and, in the event of successful commercialization, they may generate benefits to communities and other stakeholders. The probability and eventuality for such benefits therefore raise the question of whether opposing biodiversity related patents for the sake of stopping bio-piracy is the realistic approach to be adopted by any patent system. From our perspective, blind opposition to biodiversity-related patents is not a realistic approach, and the way forward is to link patents with benefit sharing. As suggested by several studies and policy initiatives such as the South African policy approach for the protection of indigenous knowledge through IP systems, perhaps co-ownership of such patents between rural communities and business operators may signify a real effort by the latter group towards ensuring 'fairness' and 'equity' in sharing the benefits arising from the utilization of genetic resources and TK.[23] There is another approach to compensating local and indigenous communities in connection with the exploitation of an alleged bio-piracy patent. This approach contends that rather than revoking such an alleged bio-piracy patent upon the request of local and indigenous communities or another actor, such as a network of environmental lawyers, the best approach could be the transfer of patent rights to the people who are offended by this patent. Advocates of this approach argue that for a patent that could be worked successfully and profitably, there is no point revoking it because nobody gains from such a revocation.[24]

A critical stage during the consultation process is the negotiation of a benefit-sharing agreement between users and providers of community assets. To demonstrate their real commitment to ensuring that communities are provided with real benefits, this agreement should omit clauses such as 'benefits to rural communities are royalties deriving from successful development of commercial products'. Instead, like other counterparts, the communities should also benefit directly from the overall project, whether or not there is successful development of commercial products. This proposition is based on the realization that through licensing mechanisms, research institutions and industries can make direct benefits from their patents even before the end of the development of final products. In the case of the South African *Hoodia,* the production of semi-synthetic versions of the P57 is an avenue that was investigated by the British pharmaceutical company Phytopharm to make income,[25] before the end of the development and eventual commercialization of anti-obesity pharmaceuticals based on P57, which the San communities are still awaiting in order to earn[26] any royalties as provided under their 2001 benefit-sharing agreement with the South African CSIR.[27]

Inserting the approach outlined here in a patent regime of a country rich in biological and cultural diversity will ensure that communities are enabled to express their views in processes leading to the acquisition of patents based on their assets and that they reap benefits as the right holders of their assets. These measures will also undermine, rather than eliminate, the pursuit of bio-piracy patents by unscrupulous scientific and commercial actors.

PIC and BIS in India

The 1992 Rio CBD affirms the sovereignty of member states over the biological resources found within their territories, thus securing their rights to decide upon any initiative involving such resources. Article 15(5) of the CBD provides that access to the resources shall be subject to the PIC of the contracting party providing the resources unless otherwise determined by that party. PIC is a critical aspect of access regulations and must therefore be provided in national biodiversity regulatory processes, especially in the context of the permitting scheme through which research and plant-collection permits are granted. Furthermore, the free and prior informed consent of communities is a human rights issue as discussed earlier in Chapter 4. Despite the highly centralized approach of the CBD vis-á-vis the PIC of contracting parties, the 2002 Bonn Guidelines on ABS suggests an approach which may be employed in accommodating the requirements for PIC of other stakeholders such as local and indigenous communities in access regulations. For one thing, Article 26(d) of the Bonn Guidelines provides that in addition to the consent of the relevant national authority in the provider country, seeking the consent of other stakeholders such as indigenous and local communities amounts to the fulfilment of the basic principle of a PIC. Based on this principle, Article 31 of the Bonn Guidelines emphasizes local and indigenous communities as one category of the authorities whose consent must be sought in access processes. Having local and indigenous communities as a recognized authority whose consent must be sought is, according to the Bonn Guidelines, respectful to their established rights associated with the genetic resources and TK being accessed. However, despite this internationally agreed approach of the Bonn Guidelines, it is the case that in exercising its sovereign rights over access to its resources, a country like India appears to have concentrated all the decision-making powers in the hands of designated bodies such as the NBA and State Biodiversity Boards, as the sole bodies whose PIC is legally captured under its access regulations. This may not be surprising because the CBD is a legally binding international environmental treaty while the 'Bonn Guidelines' is very much a voluntary instrument that CBD member states are not obliged to comply with. But considering the important role that local communities play in the development, conservation and sustainable use of India's wealth of biological and genetic resources and TK and practices, this country must consider a proper incorporation of the PIC of its local and indigenous communities in access processes that target their assets.

As hinted above, the regulatory framework in India over biodiversity issues fails to provide for the PIC of local and indigenous communities within the framework of access processes. This situation is obvious in one important piece of legislation in India, notably the Biodiversity Act 2002. Section 3 of the Biodiversity Act 2002 dealing with regulations of access to biological diversity stipulates that access to India's biological diversity is subject to the approval of the NBA for people who are not citizens of India; or to citizens of India who are non-resident in India; or to corporate bodies, associations or organizations that are not incorporated or registered in India or those registered in India under any law for the time being in force which has any non-Indian participation in its share capital or management. The most important point in this provision is that the NBA is the sole body legally responsible for granting access authorizations to actors who apply for them. Chapter III of the Biodiversity Act 2002 deals with the NBA. Section 8(4) of the Biodiversity Act 2002 provides for the composition of the NBA which omits participation of local and indigenous communities. Another important institution provided by the Biodiversity Act 2002 is the establishment of the State Biodiversity Board, which has the power to allow or restrict Indian citizens or India's registered corporate bodies and associations to undertake the collection of India's biological resources. The State Biodiversity Board's decision to restrict such collections should only occur if they appear to be detrimental or contrary to the objectives of conservation and sustainable use of biodiversity. So far, what appears obvious in the Indian legal context is the exclusion of local communities from the regulatory processes pertaining to access and thus a severe disrespect to their human rights of free and prior informed consent and participation in the relevant processes. However, in what may seem to be a catching-up element of the Act in respect of the rights of local communities, Section 41 provides for the establishment of biodiversity management committees. The Act stipulates that the purpose for creating these committees in local areas is for the promotion of conservation, sustainable use and documentation of biological diversity, and the chronicling of knowledge relating to biological diversity. Importantly, the above-mentioned bodies (the NBA and the State Biodiversity Board) are required to consult the biodiversity management committees when making any decisions relating to the use of biological resources and knowledge associated with such resources occurring within the territorial jurisdiction of the biodiversity management committee.

From the perspective of this work, the regulatory framework for access to biological resources in India does not accommodate the mechanism for the protection of the rights of prior and free informed consent of local and indigenous communities. The requirement that the relevant bodies including the NBA and the State Biodiversity Board must consult local communities when making decisions about access and eventual use of their assets cannot be approximated to proper compliance with the PIC and BIS principles. Indeed, all access applications are submitted to and handled by the approval bodies. When consulting with the local management committees, there is no guarantee that these bodies will share all information with the management committees in a form that enlightens the latter sufficiently about the project and activities at stake. On the other hand, the existing framework regulating access to biological resources in India does not require any interaction between outsiders seeking to access and acquire community biological assets and the communities themselves. Thus, the current system fails to establish a platform for consultations between outsiders and the communities. If such a platform was established, it could be used by outsiders to seek the PIC of local and indigenous communities. Furthermore, it is nowhere legally established that when consulting the biodiversity management committees, their views can play any role in influencing any decisions in respect of the granting or approval of authorizations, decisions that are made by the relevant bodies.

To make the much-needed amendments to the system thus incorporating community rights of PIC in access regulations in India, the regulations must provide for a platform of interaction between local and indigenous communities and applicants for access permits. In addition, there should be another platform of interaction and exchange between local and indigenous communities and the relevant approval bodies. All these platforms must operate in a way that ensures full disclosure of the intent of access applicants to local communities in respect of their proposed activities. Ideally, all stakeholders must have the same level of knowledge and understanding of the purpose of the activities for which access permits are sought, practically balancing the amount and type of information known by everyone, thus fulfilling the BIS principle. The realization of the right to prior and free informed consent of local and indigenous communities, through compliance with the mechanism contemplated here, will also help in the realization of community right to self-determination as provided by regional and international human-rights instruments and considered by the works of regional human-rights settings such as the African Commission of Human and Peoples Rights. This is because, while local and indigenous communities will have their rights of free and prior informed consent included in the access regulations, they do not have the power to approve or issue a plant collection or research permit to an applicant. Their involvement in the realization of the participatory approach of access regulations will not trespass upon the responsibilities of the legally recognized decision-making bodies; neither will it have any affect on the territorial integrity of India.

Seeking the consent of rural communities to use the PGRs and associated TK can be a tricky undertaking, especially as the knowledge can be accessed in ethno-botanical publications if it has already been subject to ethno-botanical research. For example, before going to some specific communities, it is very likely that a researcher would have already read about their knowledge in a publication. It is my view that if the knowledge of rural communities is likely to be instrumental in any research, whether or not such knowledge is published, the PIC of the relevant communities should be sought prior to any utilization of their knowledge. However, another aspect of the difficulties involved is that the knowledge can also be spread across communities in different areas. As this is commonly the case, the researcher should have the responsibility to undertake a thorough census of all the communities sharing such knowledge and negotiate with their representatives about potential commercial or scientific use. As indicated earlier, the first communities contacted will be very useful in this activity as well as the assistance of experts, such as anthropologists. Throughout such negotiations, the benefit-sharing mechanism will be discussed and, eventually, communities' benefits should be put in a local trust fund that will be administered by representatives of the communities themselves. Overall, the PIC of the communities is essential because it empowers them to say what their expectations are regarding the way research based on their assets should be conducted.

Enforcement and wider applicability of the regulatory measures

National biodiversity regulatory measures should provide for effective enforcement mechanisms, empowering local communities to initiate and pursue legal challenges against actors who trespass upon their rights in the course of access and potential use of their assets. Access to justice by local and indigenous communities through compliance with the traditional judicial proceedings is therefore critical for effective implementation of biodiversity regulatory measures aimed at the protection of the rights of local and indigenous communities. Provided other public services including the police, forestry administrations and the judiciary work and cooperate comprehensively based on a less corrupt and more transparent ethos, national enforcement of biodiversity regulations should not be a complicated undertaking and will just act as a deterrent to those actors who willingly or mistakenly might be embarking upon unlawful practices in respect of the exploitation of community assets. The trickiest issue regarding the implementation of national biodiversity measures is their enforceability beyond national borders, for instance in order to address a case such as the hypothetical one described in the previous chapter. Here we attempt to apply a variety of tools in a bid to challenge any actor based overseas and to hopefully bring about compliance with developing countries' biodiversity regulatory measures by actors operating beyond national borders.

The hypothetical case: role of private international law on the demands of the plaintiffs

The enforcement of the regulatory measures discussed here is likely to be very contentious when overseas actors are involved in a case such as the hypothetical one presented previously. Despite, as presented below, the obvious constraints associated with the enforceability of the biodiversity regulatory measures beyond national borders of biodiversity-rich countries, there is room for opportunity that the plaintiffs could exploit in a bid to earn the very minimum satisfaction in respect of overseas partners' compliance with national regulatory approaches on access to and collection of the community assets from developing countries. There are parameters that developing countries' actors (the plaintiffs in the context of our hypothetical case) can use to measure their satisfaction from a judicial proceeding overseas. For example, the plaintiffs' satisfaction from this case may be measured if they succeed in getting the judicial process in Japan to establish jurisdiction over this case in Japan. An additional bonus to such satisfaction will emerge if the established jurisdiction makes a judgement which meets the plaintiffs' expectations (such as the highly unlikely revocation of a patent). Another no less important measurement of satisfaction from the perspective of the plaintiffs in our hypothetical case is the acquisition of compensation from the (supposed) Japanese offender in this case. Let us now address the possible outcomes that one could expect from the application of private international law in Japan aimed at the establishment of jurisdiction over the hypothetical case and the possibility for revocation of any patent involved in this case. Thereafter, we will turn to discuss the route of compensation to be granted to the plaintiffs if revocation of the patent appears to be a difficult course of action.

Theoretically, the interpretation of the *Jori* concept and the Savigny model in relation to our hypothetical case suggests that the plaintiffs can validly bring Team B to justice in Japan. The judge can establish *Jori* over the case on the grounds that Team B is based and conducts its research and prospective commercial partnerships in Japan. The trouble is that the court in Japan can also establish special circumstances related to the case, which can make it inappropriate to be heard in Japan. Efforts to establish jurisdiction over our hypothetical case based on the plain and theoretical interpretation and application of *Jori* and/or the jurisdictional grounds of the CCPs should recognize the special nature of the case. This case is about a dispute arising from the use of biological elements over which traditional rights are claimed and are legally protected by the IP-based and human-rights oriented biodiversity regulations of a developing country D, but not protected under any law in Japan. Basically, assuming that the said regulations are actually in force in Country D, the plaintiffs must be able to pursue judicial challenges against those who infringe their rights within the national boundaries. However, if the plaintiffs succeeded in pursuing such legal actions in Japan, this would mean that the Japanese approach in the choice of applicable law is flexible because it might have allowed Country D's regulations to benefit from positive consideration through the judicial process in Japan, especially in the absence of an enforceable international legally binding treaty accommodating the issues covered by Country D's regulations to which Country D and Japan should be member parties. But, as we describe below, practically speaking, acts committed by a Japanese team considered as infringements to the rights of Country D's rural communities according to its IP-based and human-rights-oriented regulations, are lawful and may be in the public interest in Japan if they are not specifically prohibited under the Japanese law.

Let us consider as 'tortious act' the alleged infringements of the rights of rural communities through the acts committed by Team B. Under the Japanese legal system, the courts accept tort as a jurisdictional act if the place where it was committed and produced damages was in Japan.[28] In relation to our hypothetical case, if the alleged infringing activities occur in Japan, the purported damages to the communities do not occur in Japan. It is the approach of the Japanese legal system that, unless a particular subject matter is protected under the Japanese law, exploiting or using this subject matter in any way, including for scientific and commercial purposes, cannot be considered a tortious act. Thus, in Japan the manufacturing and sales of goods protected by foreign law and not protected in Japan is a lawful activity. This is an implication of the principle of territoriality, which appears in The Hague Convention on Choice of Court Agreements 2005 (hereinafter 'The Hague Convention').[29] However, according to experts such as Professor Toshiyuki Kono of Kyushu University School of Law, even if a foreign law was applicable in Japan, for a court to establish and have jurisdiction over a given case, the plaintiff must provide evidence of special and additional circumstances in relation to the alleged infringement. Acceptable additional circumstances include proof of injury, such as huge economic loss caused by the tortious activities. In our hypothetical case, the lawsuit filed by the plaintiffs will hardly lead to favourable judgements for the latter because, if PGRs and TK are protected under Country D's regulations, such regulations cannot be applicable in Japan. In this respect, the activities of Team B and eventual commercial partners are lawful in Japan. It is therefore difficult for the plaintiffs to prove that a tortious act has been committed. Nonetheless, the plaintiffs may consider giving proof of additional circumstances related to the case, raising the damages caused by the alleged infringement committed by Team B. In relation to our hypothetical case, it is likely that as damages, the plaintiffs will mention the negative impacts that Team B's activities might have had on the spiritual and cultural dimensions of the lifestyles of local communities in Country D. For example, the communities may state that the fact that Team B filed patents over the derivatives of the plants that they supplied to Team A, as it is a strong indication of commercial intent, is harmful to their culture and was not authorized by them in the first place. Such additional circumstances are unlikely to be heard in Japan and cannot justify the establishment of jurisdiction over our hypothetical case in the context of the Japanese legal system.

Considering the constraints associated with the establishment of jurisdiction over this case in Japan and thus in achieving any revocation of the patent as requested by the plaintiffs, the latter may adopt a realistic approach and request compensation from the benefits generated by the Japanese counterpart. However, can compensation be allocated to the plaintiffs based on Country D's regulations and the agreement between Team B and Team A? We address this issue by exploring the possibility for including a benefit-sharing clause in the contract. The Japanese approach in the choice of applicable law in contractual negotiations is regulated under Article 9 of Horei Application of Laws (General) Act No. 10 of 1898, as amended by the 1986 Act No. 84. Article 9(2) of Horei provides that: 'The creation and effect of a contract are governed by the place where the offer is made.' If the offeree at the time of acceptance does not know where the offer was made, the offerer's domicile is deemed to be the place of acting. Moreover, Article 4(1) of Horei Act No. 10 of 1898 provides that 'the law of the country of a person's nationality governs the grounds upon which he may be adjudged incompetent, and the law of the country of adjudication governs the effect of adjudication'. In respect of our case, if the contract signed between the plaintiffs and the defendant did not provide for any choice of applicable law, the two articles of Horei mentioned above render the case complicated for the plaintiffs, suggesting that the normal rules of choice of law under international private law are the alternative.

The offerer of the partnership in this case is the defendant whose place of domicile is Japan. The applicable law will be the Japanese law to which the issue is closely connected. Consequently, there is very little room for the plaintiffs to expect any positive outcome in terms of being compensated as required under their country's national regulations. Let us assume that Country D's regulations could be chosen as applicable rules in the determination of compensation because the negotiations of the contract started in Country D. However, although they begun in Country D, negotiations continued and completed through emails and faxes, after the representatives of Team B had flown back to Japan. The determination of the place of contract for the purpose of choice of the applicable law is very difficult in these circumstances. If all parties in conflict agreed that Country D is the place where the contract was made, theoretically, its regulations would be chosen as the applicable law. However, this approach can be strongly rebutted by the Japanese court on the grounds of the territoriality of Country D's regulations, which cannot be applied in Japan. Second, agreeing that Country D is the place where the contract was made can prove difficult. In objection to this position, the defendant can argue that although the negotiations of the contract started there, they were subsequently conducted via emails and faxes, and perhaps the representatives of Team A also travelled to Japan for such negotiations. With these elements, the Japanese court is unlikely to adopt the compensation schemes according to Country D's regulations as required and expected by the plaintiffs.

The Route of MoUs among countries involved in ABS transactions

Nobody knows what will be the form, shape and legal status of the international regime on ABS currently under discussion by member parties to the 1992 CBD and expected to complete by the tenth meeting of the CBD COP scheduled in Nagoya, Japan in October 2010. Even if these discussions culminated in a treaty which is legally binding, it is still not clear how much consideration the rights of local and indigenous communities would receive and whether this regime would bear a strong compliance mechanism backed by severe sanctions such as commercial retaliation to non-compliant countries. In the context of an international treaty, Peter Drahos argues for the setting up an international enforcement pyramid. He suggests the creation of a multilateral entity called the global bio-collecting society (GBS), which will play a central role in this pyramid. The GBS could guide indigenous groups, states and the civil society on how to use the pyramid.[30] However, it is obvious that in the absence of an international and perhaps a legally binding treaty on genetic resources and TK, the wider applicability of the measures discussed in this book will not be easy if they are adopted by developing countries rich in biological and cultural diversity. National regulatory measures will always tend to reflect national realities and address national interests. In this respect, getting other countries to adopt developing countries' standards (such as those of Country D) may perhaps impact upon other states' policy objectives and sovereignty. However, getting other countries to accommodate these measures in their national biodiversity policies is not completely impossible. One way towards ensuring that these options are applied by other countries is through the development of MoU between a developing country such as Country D in our hypothetical case and other countries.[31] Let us stress on our hypothetical scenario of exploitation of PGRs and TK in Country D by a foreign operator based in Japan. Country D can take advantage of such activity and establish a MoU with the country seeking access (e.g., Japan). According to Drahos, 'the flexibilities of MoU would be of advantage in an area where States would have to find ways of agreeing to the enforcement of different standards of protection'.[32]

During the negotiations of the MoU, the major challenge to Country D's negotiators will be to ensure that key elements of their national regulatory measures are taken into account in the MoU, and this will require some strong bargaining skills. If therefore through these negotiations Country D was able to use the MoU to stimulate compliance with their regulatory standards by other countries, arguably the other countries will gradually see the need to establish their own biodiversity regulatory measures stressing the PIC and BIS, improved patent approaches, the benefit-sharing and human-rights-related concerns of local and indigenous communities recognized in international and regional human-rights instruments. In the long run, such measures would ensure that in other countries patents are not granted on products or processes derived from the exploitation of developing countries' resources, unless the patent examiner is satisfied with the report of the PIC process and proof of the disclosure of origin. Therefore, provided that Country D or our study countries adopt the suggested options, they will need to critically investigate the possibilities of using MoU in a bid to promote the wider applicability of their schemes.

The route of arbitration and the New York Convention 1958

There is another route that might provide room for the allocation of compensation to the plaintiffs, either as provided under Country D's regulations or as provided under a neutral regulatory framework, thus arguably allowing such regulations to have some effects in Japan. This seems to be feasible by including a benefit-sharing or compensation clause in the contract, which needs to identify compensation as arbitral awards. The 1958 New York Convention[33] provides a legal framework for this approach. Article II of the 1958 New York Convention requires each member state to recognize an 'agreement in writing' in which the parties undertake to submit to arbitration. The term 'agreement in writing' is defined as an arbitral clause in a contract. The arbitral clause must require the court of a contracting state to refer parties to arbitration when seized of an action in respect of which the parties have made agreement as defined by the convention. Furthermore, Article III of the New York Convention requires contracting states to recognize arbitral awards as binding and to enforce them 'according to the rules of procedure of the territory where the award is relied upon'. Moreover,

Article IV of the New York Convention outlines the conditions for the recognition and enforcement of such awards. It provides that in order to be recognized and enforced by the contracting parties, arbitral awards must reflect the provisions of Article I(2) of the New York Convention, which stipulates that 'arbitral awards' are awards made by the arbitrators appointed for each case or awards made by permanent arbitral bodies to which the parties have submitted.

Concerning our hypothetical case, if the parties to the contract had decided to follow the arbitration route in the event of conflicts, they should have appointed an arbitrator or submitted their contract to a permanent arbitral body. Such appointment, to be stipulated in the arbitration clause of the contract, must also clarify that the decision of the arbitrators about issues such as compensation is the 'arbitral awards'. Arguably, this way, compensation is an 'arbitral award' in the sense of Article I(2) of the New York Convention. If as a result of the arbitration process the compensation awarded to the plaintiffs reflects the compensation scheme envisaged in the regulatory measures of Country D, as a member state of the New York Convention, Japan must recognize and enforce those awards. The WIPO Mediation and Arbitration Centre[34] can be chosen by contractual parties as the independent arbitration body. The centre deals with cases of partnerships in research and development and other IP cases related to patents, trademarks and domain names. Our hypothetical case has a strong R&D component and therefore falls within the ambit of the centre's areas of interests. Through the process, the independent arbitration body could apply neutral rules, such as the general principles of law, including the International Institute for the Unification of Private Law (UNIDROIT) principles of international commercial contracts 2004.[35] In the event of dispute, the centre will assist the parties in the constitution of the three-member arbitration tribunal, one arbitrator chosen by each party and the third by chosen by the first two arbitrators. However, the centre can also suggest an arbitral tribunal to the conflicting parties if they so wish. If the judgement of the arbitral tribunal indicates that compensation should be given to the plaintiffs as stipulated in the contract, arguably the mechanism provided by Country D's regulations will be used as the framework to guide its allocation.

The recognition of compensation as arbitral awards will not go without difficulties. Article III of the 1958 New York Convention provides that the recognition and enforcement of arbitral awards are subject to 'rules of procedures of the territory the award is relied upon'. Furthermore, Article V of the 1958 New York Convention describes the grounds upon which the recognition and enforcement of the awards may be refused at the request of the party against whom it is invoked. Article V(a) provides that the recognition and enforcement of foreign arbitral awards can be refused if the parties against whom it is invoked provide evidence that the parties to the 'agreement'[36] were not under the law applicable to them, were negotiating with some incapacity, or the said agreement is not valid under the law to which the parties have subjected it, or, failing any indication thereon, under the law of the country where the award was made. Arguably, these prerequisites suggest that in relation to our hypothetical case, the defendant has sufficient room to oppose the judgement of the arbitral tribunal, making it difficult to allocate compensation as 'arbitral awards' to the plaintiffs which the latter expects to reflect the approach provided by its national regulations.

Conclusion

Conceptualizing biodiversity regulations that are IP-based and human-rights oriented, especially addressing the concerns of local and indigenous communities in developing countries, is one thing. Putting such theoretically conceptualized measures into practice is another, but it is certainly not impossible. This chapter has discussed how IP-based and human-rights-oriented biodiversity regulations can work in practice at the national level to ensure community participation in the decision-making processes; to ensure that in participating, their community PIC is sought by other actors and to ensure that patent processes are considerate of community human-rights concerns. Unsurprisingly, the enforceability of these measures is not straightforward when they are allegedly infringed by an overseas actor. Despite any constraint in their enforcement, developing countries rich in biological and cultural diversity should take the lead in promoting respect of the rights of their forest-dwelling communities over those resources, because, only then, can these communities be encouraged to commit seriously to conservation and sustainability of the resources and to pursue their flexible ways of sharing.

9 Final remarks

This work discussed the fundamental criteria for the patentability of inventions and for the protection of new plant varieties to the extent that they are based on the exploitation of community assets, which are defined in this work as PGRs and associated TK, innovations and practices of local communities. IP regimes work to provide IP-holders with legal rights over the outputs of their inventive undertakings, the products of their creative minds. In this respect, and provided inventors fully complied with the relevant basic and fundamantal criteria, patents and PBRs granted to them can be viewed as working the way they are supposed to, that is, to protect scientifically proven inventions that comply with the fundamental criteria for patents and PBRs. Any other thing such as the TK and practices of local communities and the plant materials to which such knowledge is associated, as they are less likely to be grounded on modern scientific methodologies and approaches, will fail the patentability and PBRs criteria, and thus will not be legally protected. Patents and PBRs do not work for community assets and community rights over their assets. On their part, local and indigenous communities hold culturally rooted customary rights over their assets, which they use and manage according to their traditional systems. In line with their traditional systems of use and management, local and indigenous communities demand rightful control over access to and use of their assets by others. However, unlike IPRs such as patents and PBRs which can limit access to the protected subject matters, community rights over traditional assets are exercised in a flexible manner and do not fundamentally prevent non-community members to access and use community assets. These two rights systems are thus fundamentally different. Based on this difference, community-rights holders and actors sympathetic to community concerns understandably complain about issues such as bio-piracy and disrespect to critical elements of community rights by scientific and commercial actors, including but not limited to such elements as the rights of PIC, benefit sharing and participation in decision-making.

The IP-based and human-rights-oriented regulatory approaches described and discussed in this book are meant to bring about harmony and understanding between the two systems of rights. Their adoption and implementation by developing countries rich in biological and cultural diversity will help local communities and national actors understand what IPR systems are all about, especially when applied to nationally and legally protected community assets. This book proposes approaches that contemplate the setting up of platforms of discussions and consultations among various stakeholders involved in access operations at the national level, making it possible to improve understanding and establish trust. Their adoption and implementation at the national level will also empower local communities to pursue legal challenges against stakeholders who trespass upon their legally recognized rights. Moreover, despite the many difficulties that are likely to be faced in an attempt to pursue the international enforceability of the proposed regulatory mechanisms, possible routes for their wider applicability are suggested. While the development and eventual implementation of the suggested regulatory mechanisms are obviously not easy in practice, this study finds that they are not completely impossible and can be taken up by countries at the very least at the national level.

Appendix I

CITES permitting approach in brief

The Convention on International Trade in Endangered Species of Wild Fauna and Flora (CITES) was concluded at a meeting of 80 countries on 3 March 1973 in Washington, DC, in the USA and entered into force on 1 July 1975. Otherwise known as the Washington Treaty, CITES is concerned with threats to the sustainability of valuable species of wild fauna and flora, which are caused by the international trade in such species. Because international trade in wild species of fauna and flora means the movement of specimens across borders from one country to the other, there is essential need for international cooperation among countries in order to control trade in wildlife specimens of plants and animal for the purpose of achieving conservation and sustainability. Today, CITES membership extends to 174 countries, which, at least theoretically, is an indication of the seriousness of the international community on issues such as the depletion of wildlife species caused by international trade activities on them. CITES carries out its activities through various organs including the Conference of the Parties, which is the supreme decision-making body of the Convention. Members gather every two years to review progress on the implementation of the Convention and to decide upon issues such as the listing or delisting of species in the various appendices, according to their changing level of threats. Other organs of CITES include the Standing Committee, which provides guidance to the Secretariat regarding the implementation of the Convention as well as overseeing the management of the Secretariat's budget, the Plant and Animal Committees, and the Nomenclature Committee, all of which are seen as the technical bodies of the Convention. The Convention's day-to-day activities are run by the Secretariat, which plays a pivotal and fundamental role in its implementation. Each of these bodies is entrusted with specific responsibilities either as provided by the Convention or according to the resolutions of the Conference of the Parties.

There are more than 30,000 species of wild fauna and flora to which CITES accords varying degrees of protection. The way CITES rules control trade in these species is by organizing them in categories called appendices and by establishing a licensing scheme that parties may follow in authorizing exploitation and exportation of these species for commercial purposes. Each member party to CITES must set up a management authority which is essentially responsible for issuing the relevant permits, without which exportation for trade cannot be carried out; and the scientific authority, which essentially deals with the technical aspect of conservation of the species, and must therefore manage scientific data that underpins the categorization of the species.

CITES classifies species in three categories according to the level of protection that they need. Species are categorized in three appendices, notably, Appendix I, Appendix II, and Appendix III.

- Appendix I cover species which are threatened with extinction. Trade of specimens listed in this appendix is permitted only in exceptional circumstances. Trade in specimens will require depending on a specific case, certificates such as an import permit, an export permit and a re-export certificate, all of which must be processed and issued by the management authority
- Appendix II covers species which may not necessarily be threatened with extinction, but for which trade must be controlled in order to avoid utilization incompatible with the survival of the species. Export permits and re-export certificates are needed for trade in these species, all to be granted by the management authority of the country of export. There is no need for an import permit for Appendix II species.
- Appendix III covers species that are protected in at least one country which has requested assistance from a fellow member of CITES in controlling trade in these species. Each party is therefore entitled to make unilateral amendments to this Appendix. An actor wanting to trade from a state that included species in this appendix must acquire an export permit to be issued by the management authority of this state. If the species has to be exported or re-exported from another state, an export permit or a re-export certificate must be acquired from the management authority of this state.

It must however be stressed that for trade in any CITES-listed species to be permitted based on the permitting mechanism of the 1973 Convention, these species must have been exploited, collected or harvested legally beforehand, according to national legislation on access to and collection of plant or animal resources in force in the country.

Appendix II

Laws, policies, professional and ethical codes and declarations

Universal Declaration of Human Rights (1948)
ICCPR (1966)
International Covenant on Economic, Social and Cultural Rights (1966)
The UN-DRIP Resolution 61/295 of the 61st General Assembly of the United Nations (New York, 2007)
United Nations CBD (Rio de Janeiro, 1992)
Agreement on TRIPS of the WTO (1994)
The Conventions of the UPOV (1961, 1978 and 1991)
Convention on the Grant of European Patent (EPC) (1973)
ILO (No. 169) Concerning Indigenous and Tribal Peoples in Independent Countries (1989)
FAO International Undertaking on Plant Genetic Resources (1983)
FAO IT-PGRFA (2001)
ILO Convention No. 107 on Indigenous and Tribal Populations (1957)

Emerging international instruments

The Protection of Traditional Knowledge: Draft Objectives and Principles. Document Prepared by the Secretariat, WIPO (WIPO), 2006, WIPO/GRTKF/IC/10/5
The Protection of Traditional Cultural Expressions/Expressions of Folklore: Draft Objectives and Principles, Document Prepared by the Secretariat, WIPO (WIPO), 2006, WIPO/GRTKF/IC/10/4
The International Regime on ABS is being discussed under the auspices of the CBD Working Group Ad hoc Open Ended Working Group on ABS. See last Draft version of the International Regime annexed to the Report of the 7th Meeting of the Working Group Held in Paris 2 to 8 April 2009. UNEP/CBD/WG-ABS/7/8, p. 21
OAS Draft Resolution on the Draft American Declaration on the Rights of Indigenous Peoples (Washington, DC, 2008)

National laws and policies

United States Patent Act 35 in the United States Constitution (1790)
Japan Patent Act No. 121 of 1959
The Biodiversity Law No. 7788 of 30 April 1998, Official Gazette No. 101, 27 May 1998 (Costa Rica)
Law No. 20 of June 26, 2000: Special Intellectual Property Regime Governing the Collective Rights of Indigenous Peoples, for the Protection and Defence of their Cultural Identity and their TK and Other Provisions (Panama)
Law No. 27811 Introducing a Protection Regime for the Collective Knowledge of Indigenous Peoples derived from Biological Resources, Official Journal, EL PERUANO, 10 August 2002 (Peru)
Provisional Measure on 'Access to Genetic Resources and Traditional Knowledge', No. 2186–16, of 23 August 2001 (Brazil)
The Biological Diversity Act, 2002, along with the United Nations CBD and the Johannesburg Declaration on Sustainable Development (Rio + 10). Universal Law Publishing, Delhi (India)
Ethiopian ABS Law: Proclamation No. 482/2006 of 27 February Entitled 'A proclamation to Provide for Access to Genetic Resources and Community Knowledge and Community Rights'
Kenyan ABS Law: Legal Notice No. 160 the Environmental Management and Coordination (Conservation of Biological Diversity and Resources, Access to Genetic Resources and Benefit Sharing) Regulations, 2006
Peruvian Ley No. 26839, which is the Legislation on the Conservation and Sustainable use of Biological Resources (1997)
South African National Environmental Management Biodiversity Act No. 10 of 2004
National Environmental Management: Biodiversity Act (10/2004): Commencement of Bio-Prospecting, ABS Regulations, Government Gazette, Republic of South Africa, Vol. 512, No. 30739, Pretoria, 8 February 2008
The Protection of Plant Varieties and Farmers' Rights Act 2001 (India)
The Scheduled Tribes and Other Traditional Forest Dwellers (Recognition of Forest Rights) Act 2006 (India)
Japanese Revised CCPs Law No. 109 of 26 June 1996
National Environmental Management Act 1998 (South Africa)
The South African Patent Act No. 57 of 1978
The South African Intellectual Property Amendment Act No. 38 of 1997
The South African Patent Amendment Act 20 of 2005
The South African Indigenous Knowledge Policy of 2004
The Brazilian Industrial Property Law 14/05/1996 No. 9.279
Japanese Horei-Application of laws (general) Act No. 10 of 1898, as amended by the 1986 Act No. 84
The Indian Biological Diversity Act 2002

Regional and international laws and policies

The Convention on the Recognition and Enforcement of Foreign Arbitral Awards (New York, 1958)
The Hague Convention on Choice of Court Agreement, which is not yet in force (The Hague, June 2005)
Andean Community Decision 391: Common Regime Governing Access to Genetic Resources (1996)
ACHPR (Nairobi, 1981)
Andean Community: 2001 Machu Picchu Declaration 524 on Democracy, the Rights of Indigenous Peoples and the War against Poverty (Lima-Machu Picchu, 2001)
Andean Community Decision 674 on the Consultative Council of the Andean Community Indigenous Peoples (New York, 2007)
Andean Community Charter for the Promotion and Protection of Human Rights (Guayaquil-Ecuador, 2002)
OAS Charter (Bogota, 1948)
American Declaration of the Rights and Duties of Man (Bogota, 1948)
The Inter-American Charter of Social Guarantees (Bogota, 1948)
The Inter-American Convention on Human Rights (San Jose, Costa Rica, 1969)
The Proposed American Declaration on the Rights of Indigenous Peoples (Lima, Peru, 1997)
The African Model Law for the Protection of the Rights of Local Communities, Farmers and Breeders and for the Regulation of Access to Biological Resources (Addis Ababa, Ethiopia, 2000)
Convention on International Trade in Endangered Species of Wild Fauna and Flora (Washington, 1973)
Paris Convention for the Protection of Industrial Property (Paris, 1883)
The Bonn Guidelines on ABS (The Hague, Netherlands, 2002)

Professional and ethical codes and declarations

The 15 international declarations and professional codes of practices surveyed by Graham Dutfield in 2002 are:

1. The Declaration of Principles of the World Council of Indigenous Peoples (1984)
2. The Kari-Oca Declaration and the Indigenous Peoples Earth Charter (1992)
3. The Charter of the Indigenous Tribal Peoples of the Tropical Forests (1992)
4. The UN Draft Declaration on the Rights of Indigenous Peoples (1994) (Now the 2007 UN Declaration on the Rights of Indigenous Peoples)
5. The Mataatua Declaration on Cultural and IPRs of Indigenous Peoples (1993)
6. The Julayinabul Declaration Regarding the West Tropics World Heritage Centre (1993)
7. Recommendations from the Congress, Voices of the Earth: Indigenous Peoples, New Partners, the Right to Self-determination in Practice (1993)

8. Statement from the COICA/UNDP Regional Meeting on IPRs and Biodiversity (Santa Cruz Declaration, 1994)
9. UNDP Consultation on the Protection and Conservation of Indigenous Knowledge (Sabah Declaration, 1995)
10. Final Statement from the UNDP Consultation on Indigenous Peoples' Knowledge and IPRs (Suva Declaration, 1995)
11. Principles and Guidelines for the Protection of the Heritage of Indigenous Peoples (1995)
12. International Alliance of Indigenous Tribal Peoples of the Tropical Forests: the Biodiversity Convention—the Concerns of Indigenous Peoples (1995)
13. Indigenous Peoples' Biodiversity Network: Indigenous Peoples, Indigenous Knowledge and Innovations and the CBD (1996)
14. Results of the International Meeting of Indigenous and Other Forest-Dependent Peoples on the Management, Conservation and Sustainable development of all Types of Forests: a Contribution to the Intergovernmental Panel on Forests (Leticia Statement, 1996)
15. Second International Indigenous Forum on Biodiversity: Submission to the Workshop on TK and Biological Diversity (1997)

Cusco Declaration on Access to Genetic Resources, TK and IPRs of Like Minded Megadiverse Countries (Cusco, Peru, 2002)

CPG for Participating Institutions (including botanic gardens, herbaria and museums) on Access to Genetic Resources and Benefit Sharing (Kew Botanical Gardens, London, 2001)

The Peasant's Charter: The Declaration of Principles and Program of Action of the World Conference on Agrarian Reform and Rural Development. FAO (Rome, 1981)

Appendix III
Glossary

access to genetic resources: This refers to the process of acquiring samples of biological materials or parts containing genetic material from areas within a national jurisdiction for the purpose of research on conservation, commercial or industrial application of the genetic material.

biological and cultural diversity; bio-cultural diversity: This broadly refers to a geographical region that encapsulates a sound blend of both the biological diversity as defined by the CBD and the diversity in the peoples, their languages, knowledge, innovations and technologies which are all rooted in their cultures and traditions.

biological diversity; biodiversity: This is referred in the CBD as the variability among living organisms from all sources including *inter alia,* terrestrial, marine and other aquatic ecosystems and the ecological complexes of which they are part including the diversity in species, between species and ecosystems.

biological resources: These include genetic resources, organisms or parts thereof, populations, or any other biotic component of ecosystems with actual or potential use or value for humanity.

bio-piracy: Broadly refers to the failure to acknowledge the origin of biological and genetic resources and TK, innovations and practices in an appropriate manner in the course of bio-prospecting.

bio-prospecting: Broadly refers to the searching and collecting of biological and genetics resources with or without the associated TK, innovations, practices and technologies of local and indigenous communities for the purpose of commercial development

community-based intellectual property rights: Rights held by local communities, in a communal manner, over the vital elements of their culture including traditional songs, crafts, sculptures, dances, medicinal plants, and the knowledge associated with these plants on their use at the local level.

community rights: According to the African Model Law for the Protection of the Rights of Local Communities, Farmers' Rights and for the Regulation of

Access to Biological Resources, community rights are those rights held by local communities over their biological resources or parts or derivatives thereof, and over their practices, innovations, knowledge and technologies.

cultural diversity: Broadly refers to the variety or multiformity of human populations and their social structures, belief systems, and strategies for adapting to situations in different parts of the world.

customary law: Law that is customary, that is law rooted in the customs and traditions of local communities. There have been requests that customary law may be given some recognition in the protection of TK. (See WIPO document, No. WIPO/GRTKF/K/16.)

farmers' rights: Farmers' rights mean rights arising from the past, present and future contributions of farmers in conserving, improving and making available PGRs, particularly those in the centres of origin or diversity. These rights are vested in the international community, as trustee for present and future generations of farmers, for the purpose of ensuring full benefits to farmers and supporting the continuation of their contributions (Resolution No. 5/89 FAO Conference).

genetic resources: The CBD refers to genetic resources as genetic material of actual or potential value.

human rights: Human rights refer to the basic and fundamental rights and freedoms all humans are entitled to.

indigenous peoples: According to ILO Convention No. 169, people are regarded as indigenous on account of their descent from populations who inhabited a country, or geographic region to which the country belongs, at the time of conquests or colonization or the establishment of present state boundaries, and who, irrespective of their legal status, retain some or all of their own social, economic, cultural and political institutions.

intellectual property rights: These are rights awarded by society to individuals or organizations over inventions, literary and artistic works and symbols, names, images and designs used in their commerce. IPRs give the rightholder the rights to prevent others from making unauthorized use of their property for a limited period.

local communities: According to the African Model Law for the Protection of the Rights of Local Communities, Farmers' Rights and for the Regulation of Access to Biological Resources, local communities are human populations in a distinct geographical area with ownership over their biological resources, innovations, practices, and technologies governed partially or completely by their own customs, traditions or laws.

patent rights: These are exclusive rights awarded to an inventor to prevent others from making, selling, distributing, importing or using their invention without licence or authorization for a fixed period of time. As a prerequisite to be awarded these rights, society requires that the holder of patent rights disclose the invention to the public. For an invention to be protected by patents, it has to be new but to have evolved through an inventive step or must be non-obvious to a person skilled in the art and must be industrially applicable.

plant breeders' rights: These are the rights granted to breeders of new, distinct, uniform and stable plant varieties. PBRs offer protection for at least 20 years.

plant variety protection systems: These refer to the two main treaty systems regulating the protection of new plant varieties including the agreements established under the auspices of the Union Internationale pour la Protection des Obtentions Vegetales and the TRIPS.

prior informed consent: The African Model Law for the Protection of the Rights of Local Communities, Farmers' Rights and for the Regulation of Access to Biological Resources defines PIC as the giving by a collector of complete and accurate information, and, based on that information, the prior acceptance of that collector by the government and the concerned local community or communities to collect biological resources or indigenous knowledge and technologies.

sui generis: This is a Latin expression meaning *of its own kind.* A sui generis system of protection of community rights over TK and biological resources would thus be a system of protection that takes into account the peculiarities of community rights and their valuable assets. It has to be separate from conventional IPRs.

traditional germplasm or landrace: This is a farmer's developed cultivar of crop plants that are adapted to local environmental conditions.

traditional knowledge: Article 8(j) of the CBD refers to TK as the knowledge, innovations and practices of indigenous and local communities deriving from customary use of biological resources and associated cultural practices and traditions. The general literature refers to traditional environmental or ecological knowledge as a body of knowledge and beliefs transmitted through oral tradition and first-hand observation. It includes a system of classification, a set of empirical observations about the local environment and a system of self-management that governs the use of resources.

traditional medicine: Refers to health practices, approaches, knowledge and beliefs incorporating plant, animal and mineral based medicines, spiritual therapies, manual techniques and exercises, applied singularly or in combination to treat, diagnose and prevent illnesses or maintain well-being (WHO).

Notes

1 Setting the scene

1 The concept of 'community rights' is further developed in the subsequent paragraphs of this chapter while avenues for the legal protection of elements of community rights based on a human-rights approach are suggested in the final chapters of the book.
2 I have deliberately chosen to avoid discussing the definitional issues of the terms 'indigenous', 'local', 'traditional', 'peoples', 'communities' or 'societies'. I have chosen not to do so because I consider a discussion on such issues as distractive and as tactics used in international forums by some actors to purposively divert attention from the real issues. The real issues are the need to explore opportunities and develop workable policies and legal tools that can be used to address the rights-based claims of forest-dependent peoples over their assets: plant genetic resources (PGRs), traditional germplasm and associated traditional knowledge (TK). Throughout the book and as embedded in all analysis, I will use the above-indicated terms indifferently, although some terms may be used more than others in the texts if such uses assist in clarifying the relevant arguments.
3 It is the approach of this book to define the assets of traditional communities as the biological and genetic resources and associated TK, nurtured and held by traditional communities and generally accessed from their lands and territories through various means applied by non-members of their communities to be used in scientific and commercially oriented activities. The means applied by non-members of the communities to access and acquire these resources could be, but are not limited to, plant-collection trips into the territories where these plants occur or ethnobotanical studies during which communities are interviewed about the usefulness and practical utilize of their assets in their daily life.
4 Ragavan 2001.
5 Brush 1993. See also Griffiths 1993.
6 Tonye 2004.
7 Zerbe 2005.
8 Reyes-Garcia, *et al.* 2003.
9 Professor Anil Gupta conceptualizes the development and ownership of TK by individuals in a community context. This author argues that although TK is generally communally owned at the community level, there are local experts who have developed some categories of knowledge and have kept them secret for themselves. The development of property-rights systems over TK should recognize this fact and avoids curtailing individual rights over some categories of knowledge. See Gupta 2004.
10 Kuyek 2002, p. 7.
11 Brown 2003.
12 Gibson 2005.
13 International Treaty on Plant Genetic Resources for Food and Agriculture of the United Nations Food and Agricultural Organization (Rome, 4 November 2002). Article 9(2) of

the FAO International Treaty on Plant Genetic Resources for Food and Agriculture (FAO-ITPGR) addressing farmers' rights stipulates that the contracting parties have the responsibility to realize farmers' rights according to their needs and priorities through its national legislation. Such national legislation should take measures to protect and promote farmers' rights including the protection of TK relevant to PGRs for food and agriculture, the right to equitably participate in benefit sharing arising from the utilization of PGRs for food and agriculture and the right to participate in making decisions at the national level on matters related to the conservation and sustainable use of PGRs for food and agriculture. Article 9(3) stresses that nothing in Article 9 shall be interpreted to limit any rights that farmers have to save, use, exchange and sell farm-saved seed/propagating material, subject to national law and as appropriate.

14 Article 14 of the Act of 1991 of the International Convention for the Protection of New Plant Varieties (UPOV Act 1991) addresses the scope of PBRs. Albeit with some limitations in the exercise of Plant breeders rights (PBRs) mentioned in Article 14(2), Article 14(1)(a) stipulates that acts in respect of the propagating materials that must require the authorization of the breeder include production or reproduction, conditioning for the purpose of propagation, offering for sale, selling or other marketing, exporting, importing, and stocking for any of the purposes already mentioned. Some of the privileges that were provided to farmers and researchers in the previous UPOV acts including the 1961, 1972 and 1978 acts were substantially undermined by the 1991 Act. For example, 'conditioning for the purposes of propagation' is covered by Article 14(1)(ii) of the 1991 UPOV Act. This particular provision strengthens PBRs by monitoring on-farm production and use of the harvested material. According to Biswajit Dhar, if a variety was being cultivated for consumption and the farmer suddenly decides to use harvested material for farming purposes, after conditioning, the breeder can intervene at the conditioning stage using these rights. This leaves virtually no possibility of farmers reusing seeds without the authorization of the breeder. See Dhar 2002.

15 FAO-ITPGR 2002.

16 FAO-ITPGR, Article 10(1) provides for the establishment of the multilateral system as follows: 'in the exercise of their sovereign rights, the contracting parties agree to establish a multilateral system, which is efficient, effective and transparent, both to facilitate access to PGRs for food and agriculture, and to share in a fair and equitable way, the benefits arising from the utilization of these resources, on a complementary and mutually reinforcing basis'. Articles 11(1) to 11(5) elaborate on the scope and coverage of the multilateral system. The multilateral system shall include the PGRs for food and agriculture listed in Annex I of the ITPGR and held in *ex-situ* collections of the International Agricultural Research Centers of the Consultative Group on International Agricultural Research (CGIAR).

17 FAO-ITPGR, Article 12, provides for the facilitated access and the conditions for access to PGRs for food and agriculture under the multilateral system. One of the conditions for access is clearly stipulated under Article 12(3)(d) as: 'Recipients shall not claim any IP [intellectual property] or other rights that limit the facilitated access to the PGRs for food and agriculture, or their parts or components, in the form received from the multilateral system.'

18 I acknowledge that states, the CGIAR and other international agricultural research institutions are parties to the FAO-ITPGR and exercise their sovereignty in the context of their contributions to the multilateral system. Materials contributed in the system are no longer traditional germplasm as such, as may be claimed by some communities. Indeed, it is very difficult for a given farming group or traditional community to claim ownership of the materials in the form that they enter the multilateral system. Nonetheless, the point here is precisely about that difficulty and about the fact that these materials, in one way or another, derive from some farming communities, even though there are difficulties in tracing them back to their origins. See Halewood and Nnadozie (2008).

19 International Covenant on Civil and Political Rights, GA res. 2200A (XXI), 21 UN GAOR Supp. (No. 16) at 52, UN Doc. A/6316 (1966), 999 UNTS 171, *entered into force* 23 March 1976.
20 The International Covenant on Economic, Social and Cultural Rights (New York, 16 December 1966). To date, there are 67 signatories and 159 parties to this covenant. Available online at http://www2.ohchr.org/english/bodies/ratification/3.htm (accessed 6 April 2009).
21 It is worth stressing that we have selected just a few human-rights instruments for the purpose of this book. Readers should not misunderstand our selection as an exhaustive list of existing human-rights treaties and declarations. Indeed, as discussed by Koopman (2008), human-rights instruments have proliferated and have extended their coverage over issues as diverse as indigenous and minority-people's issues, migrants' issues, women's issues, discrimination issues, children, marriage, and war.
22 World Wildlife Fund (2000).
23 Maffi 2001. See also Twarog and Kapoor 2004. In this book, the authors argue that threats to biological diversity have implications on cultural diversity.
24 Hladik *et al.* 1993.
25 Bell 1987; Wells 1995.
26 World Rainforest Movement 2005.
27 Lane 1990.
28 Clarke, Cavendish and Woote 1996.
29 Hughes 1998.
30 Heywood and Watson 1995.
31 ten Kate and Laird 1999.
32 Kloppenburg and Kleinman 1987.
33 Farnsworth *et al.* 1985.
34 Farnsworth *et al.* 1985.
35 Cragg, Newman and Snader 1997.
36 Sheldon, Balick and Laird 1997; see also Dutfield 2004.
37 Plotkin 2000.
38 On medical research, see, for example, http://www.gsk.com/ser/2001/ehs01/rep-37.html (accessed 28 February 2009).
39 Schultes 1998.
40 Reid *et al.* 1993; see Cullet 2005.
41 Mooney 2000.
42 Lewis 2000.
43 Farnsworth *et al.* 1985, see also Cragg *et al.* 1997.
44 In the case of Eli Lilly, a US pharmaceutical company that does research on the anti-diabetic uses of the Rosy periwinkle, a traditionally used medicinal plant from Madagascar, instead of isolating and characterizing the anti-diabetic substances from this plant Eli Lilly isolated and characterized anti-cancer substances.
45 See Erosion Technology and Concentration 2000.
46 ten Kate and Laird 1999.
47 ten Kate and Laird 1999.
48 Laird and ten Kate 2002.
49 See Cragg *et al.* 1997.
50 See Newman *et al.* 2003.
51 See PhRMA 2003.
52 PhRMA, 2003.
53 PhRMA, 2003.
54 PhRMA, 2003.
55 Demsetz 1967.
56 See Coase 1960.
57 Sedjo and Simpson 1995; see also Sampath 2005.
58 Sedjo and Simpson 1995.

59 Sedjo and Simpson 1995.
60 Weiner 1992.
61 Centre for International Environment London. 2002.
62 Weiner 1992.
63 Weiner 1992.
64 Griffiths 1993.
65 Griffiths 1993.
66 Morphy 1991.
67 Brokensha and Riley 1980.
68 Because the resources are owned and managed communally, decisions about access are not taken by one individual. See Field-Juma 1996.
69 The communal system of ownership and exchange strengthens local discoveries and inventions and is essential for local survival. See Tewolde 1996.
70 Zweifel 1997.
71 Blakeney 1999.
72 Ekpere 2000.
73 Ekpere 2000.
74 Kameri-Mbote 2003.
75 Article 1 of the African Model Legislation for the Protection of the Rights of Local Communities, Farmers and Breeders, and for the Regulation of Access to Biological Resources. Organization of African Unity (now African Union)/Scientific Technical and Research Commission (Ouagadougou, July 1998).
76 Dutfield 2002: 228–32.
77 Dutfield 2002.
78 WIPO 2001c.
79 Article 2(viii) of the Convention Establishing the WIPO (Stockholm, 1967).
80 WIPO 2001c.
81 Stephen Brush is among those experts who addressed the issue of commodification of knowledge especially biological knowledge within the realm of modern intellectual property rights (IPRs). He states that 'biological knowledge is treated like other types of intellectual or material property in that it can be monopolized and commodified. The establishment of property rights for biological materials followed the scientific transformation of agriculture and other industries that use biological materials' (Brush, 1993: 653).
82 Honore 1961, and Sherman and Bently 1999.
83 Mgbeoji 2006.
84 See some extracts from the International Development Research Center (IDRC), in which it is postulated that the term 'bio-piracy' is rather vague, because the indicators for bio-piracy make it difficult to actually assess how much bio-piracy is taking place on the ground. More details are available online at http://www.idrc.ca/en/ev-119999-201-1-DO_TOPIC.html (accessed 15 September 2008).

2 Patents, PBRs and community rights in international forums

1 Mgbeoji 2006.
2 Mgbeoji 2006.
3 TRIPS Agreement, General Agreement on Tariffs and Trade, Uruguay Round, WTO, Marrakech, 15 April 1994, Annex IC.
4 See, for example, Section 1(1) of the 1977 British Patent Act, which reflects Articles 52–7 of the 1973 EPC. According to Section 1(1) of the Act, a patent may be granted only if the invention is: (a) new, (b) involves an inventive step, (c) is capable of industrial application and (d) is not excluded as provided under Section 1(2) and 1(3). As for the exclusions, Section 1(2) of the patent Act stipulates that what is not

considered as inventions, for the purpose of the Act is anything which consists of (a) a discovery, scientific theory or mathematical method, (b) a literary, dramatic, musical or artistic work or any other aesthetic creation whatsoever, (c) a scheme, rule or method for performing a mental act, playing a game or doing business or a program for a computer, and (d) the presentation of information (Hart and Fazzani 2000).

5 Articles 2(1) and 2(2) and Article 6 of the Agreement Revising the Bangui Agreement of 2 March 1977 on the Creation of an African Intellectual Property Organization, Bangui, 24 February 1999.

6 Muralidharan (1997) 427–56.

7 Machlup and Penrose (1950); Merges (1995); Machlup (1958).

8 Other mechanisms such as 'technology transfer agreements' may be more appropriate in the diffusion of technical knowledge than patents. See Bessen 2005.

9 Geroski 1995.

10 Kushan 2002; also Hodge 2000.

11 The compulsory licensing scheme of Article 31 of the Trade Related Aspects of Intellectual Property Rights (TRIPS) Agreement is intended to safeguard the so-called public interest. See Gold and Lam 2003.

12 Simpson and Dutfield reviewed the grounds under which TK and biogenetic resources cannot be directly protected under patent laws. For more details, see Simpson 1997; Dutfield 1999.

13 Among the similarities, we can mention the duration of patent rights and PBRs and the nature of rights. One of the major differences is the fact that PBRs can protect discoveries while, at least theoretically, patents cannot. However, issues of similarities and/or differences between these two systems are beyond the scope of this book, as they have been extensively addressed elsewhere. See Dutfield 2008.

14 Helfer 2002.

15 Helfer 2002.

16 Dhar 2002. for comprehensive elaboration on the rights of plant breeders over essentially derived varieties (EDVS) and on the EDV concept.

18 Article 6(1)(b), UPOV 1978 and Article 6, UPOV 1991.

19 Article 6(1)(a), UPOV 1978 and Article 7, UPOV 1991.

20 Article 6(1)(c), UPOV 1978 and Article 8, UPOV 1991.

21 Article 6(1)(d), UPOV 1978 and Article 9 UPOV 1991.

22 Article 6(10)(c), UPOV 1978.

23 UPOV, Guidelines for the Tests for Distinctness, Homogeneity and Stability, Union pour la Protection des Obtentions Vegetales, 1988.

24 Halewood and Nnadozie 2008.

25 Hart and Fazzani 2000.

26 Gupta 2002; see also World Intellectual Property Organization 2003.

27 TRIPS Agreement, Article 29(1).

28 Lemley 2005, 105–28.

29 Koopman 2008.

30 Bagley 2003.

31 According to Wiser and Downs (1999), Section 102(f) provides a very solid pillar for considering TK as prior art under the United States Patent Act, whether or not such knowledge is published or held outside of the USA. This section precludes the award of a patent when the applicant did not invent the subject matter to be patented and does not have any geographical limitations such as Section 102(a) and 102(b) of the US Patent Act. See Wiser and Downs 1999.

32 McManis 1998.

33 Dutfield 2002a.

34 Article 29 of the Japan Patent Act No. 121 of 1959, which was modified by Law No. 220 of 22 December 1999; see also Bently and Sherman 2001.

35 Dutfield 2002a.

36 Using, selling, offering for sales, etc., products that have not been protected by patents but result from patented processes constitute infringements of the process patent. In the case *Monsanto Canada Inc.* v. *Schmeiser,* Roundup Ready Canola genes were found in Canola grown commercially by Schmeiser who never purchased the Roundup Ready Canola from Monsanto and never obtained a licence to plant it. Based on the examination of the word 'use' as provided by Section 42 of the Canadian Patent Act, RSC 1985, it was ruled that Schmeiser infringed Monsanto's patent because 'possession of a patented object or an object incorporating a patented feature may constitute "use" of the object and thus constitute infringement'. More on this case at: http://csc.lexum.umontreal.ca/en/2004/2004scc34/2004scc34.html (accessed 8 April 2009).
37 Convention on the Grant of European Patent, The EPC, Munich, 5 October 1973.
38 US patent No. 5,281,618 issued 24 January 1994: pesticide compositions containing neem seed and characterized by non-degrading solvent systems and high concentration of azadirachtin, the active ingredient in neem extract; US patent No. 5,001,146 issued 19 March 1991: patenting pesticide compositions containing neem seed extracts, characterized by non-degrading solvent systems; European Patent No. 5,124, 349 issued 20 October 1998: patenting an insecticide and fungicide products derived from neem seed extract and the methods of making and using the products.
39 Re-examination certificate, US patent No. 5,124,349 issued on 20 October 1998.
40 While recognizing the importance of searchable databases of TK, Dutfield is rather sceptical about its effectiveness in preventing the granting of some patents such as the one granted to the South African Council for Scientific and Industrial Research (CSIR) on the hoodia P57. See Dutfield 2002.
41 Visser 2004.
42 See note 40.
43 Dutfield 2002.
44 Convention on Biological Diversity (CBD), Decision IX/13 Article 8(j) and Related Articles, Bonn Germany, 2008, available online at http://www.cbd.int/decision/cop/?id=11656 (accessed 8 April 2009).
45 Annex to Decision V/16 on Article 8(j) and Related Articles, Nairobi, 15–26 May 2000.
46 Annex to Decision V/16.
47 Decision VII/16 of the Convention on Biological Diversity Conference of the Parties (CBD-COP) on Article 8(j) and Related Articles, Kuala Lumpur, 9–20 February 2004.
48 Decision VII/16.
49 Convention on Biological Diversity (2004) Decision Adopted by the Conference of the Parties to the Convention on Biological Diversity at its Seventh meeting, VII/28 Protected Areas, Kuala Lumpur, 9–20 and 27 February 2004, UNEP/CBD/COP/DEC/VII/28.
50 See 'Final Statement of the International Indigenous Forum on Biodiversity', 15 February 2008, Rome, Italy, available online at http://indigenousstatement.blogspot.com (accessed 9 April 2009).
51 Decision IX/12 access and benefit sharing (ABS), Ninth Meeting of the Conference of the Parties to the CBD, Bonn, Germany, 19–30 May 2008.
52 WIPO 2006c.
53 WIPO 2006c.
54 WIPO 2006a.
55 See Gibson 2008.
56 Halewood and Nnadozie, 2008.
57 Andersen 2007.

3 Brief overview on community rights in selected national regulatory instruments

1 United Nations Convention on Biological Diversity (CBD), Rio de Janeiro, 1992.
2 Cantuaria Marin 2002.

3 Dutfield 2002c.
4 Dutfield 2002c.
5 Executive Order No. 247, 1995. Prescribing Guidelines and Establishing a Regulatory Framework for the Prospecting of Biological and Genetic Resources, their By-products and Derivatives, for Scientific and Commercial Purposes, and for Other Purposes.
6 1992 CBD, Article 1.
7 1992 CBD, Article 15(1).
8 1992 CBD, Article 15(5).
9 1992 CBD, Article 15(4).
10 1992 CBD, Article 8(j).
11 1992 CBD, Article 16(5).
12 Correa 2004; See also Laird *et al.* 2003.
13 The Biodiversity Law 1998.
14 Law No. 20 2000.
15 Law No. 27811 2002.
16 Provisional Measure, No. 2186–16, 2001.
17 The Biological Diversity Act, 2002, along with the United Nations CBD and the Johannesburg Declaration on Sustainable Development (Rio + 10).
18 The Ethiopian Access and benefit saving (ABS) law is Proclamation No. 482/2006; the ABS law of kenya is Legal Notice No. 160, 2006.
19 Brazilian Provisional Measure, Articles 1 and 8; Peruvian Law No. 27811, Article 3.
20 Brazilian Provisional Measure, Article 30; Law No. 20, 2000, Chapter IV, Articles 17–23; and Law No. 27811, Articles 47–9, Articles 62–3.
21 Peruvian Law No. 27811 Article 5.
22 Chapter VII, Articles 24–8.
23 Tobin and Swiderska 2001.
24 Article 2 of the Peruvian Law, No. 27811, 2002.
25 Objective (d) of the Peruvian Law, No. 27811, 2002 stipulates that the objective of the regime is: to ensure that the use of the knowledge takes place with the PIC of the indigenous communities.
26 See http://scidev.net/en/news/peruvian-region-outlaws-biopiracy.html (accessed 9 September 2009).
27 South African Act, No. 10, 2004.
28 National Environmental Management: Biodiversity Act 10/2004, 2008.
29 Burgener 2003.
30 Burgener 2003.
31 Burgener 2003 In the four provinces of Gauteng, Kwa-Zulu Natal, Free State and North West, the departmental head office issues the permit, but in three of the other five provinces (Western Cape, Northern Cape and Mpumalanga) permits are issued by regional and districts head offices, copies of which are sent to the departmental head office, while in the remaining two provinces (Eastern Cape and the Northern Province) copies are not sent to the departmental head office.
32 See 2008 Biodiversity Regulations of South Africa, Chapter 2: Permit System for Bioprospecting and Exporting of Indigenous Biological Resources.
33 2008 Biodiversity Regulations of South Africa.
34 2008 Biodiversity Regulations of South Africa.
35 Andersen and Winge (2008).
36 Andersen and Winge 2008.
37 The Resolution on Farmers' Rights Adopted by the Governing Body of the IT-PGRFA at its Second Session (29 October–2 November 2007), 1 November 2007, held in Rome, Italy. The resolution is available online at http://www.farmersrights.org/pdf/Farmers_rights_resolution.pdf (accessed 14 April 2009).
38 See Chapter VI of The Protection of Plant Varieties and Farmers' Rights Act 2001 of India.

39 Chapter VI of The Protection of Plant Varieties and Farmers' Rights Act 2001 of India, Section 39(2).

4 Selected international and regional human-rights instruments and their provisions on community rights and IPRs

1 To date, there are 69 signatory countries and 160 countries that are parties to the International Covenant on Economic, Social and Cultural Rights. Status of ratification is available online at http://treaties.un.org/Pages/ViewDetails.aspx?src=TREATY&mtdsg_no=IV-3&chapter=4&lang=en (accessed 28 April 2009). As for the ICCPR, to date there are 72 countries signatories and 164 countries parties to this Covenant. Status of ratification is available online at http://treaties.un.org/Pages/View Details.aspx?src=TREATY&mtdsg_no=IV-4&chapter=4&lang=en (accessed 28 April 2009).

2 Indeed, as stipulated in General Comment No. 12 on the right of self-determination, this right must be promoted by state parties to the International Covenant on Civil and Political Rights (ICCPR), which should respect that right irrespective of whether a people entitled to self-determination depends on state party or not. See CCPR, General Comment No. 12: The Right to Self-Determination of Peoples (Article 1), Office of the Commissioner for Human Rights, 13 March 1984, available online at http://www.unhchr.ch/tbs/doc.nsf/(Symbol)/f3c99406d528f37fc12563ed004960b4?Opendocument (accessed 28 April 2009).

3 See Paragraphs 1 and 2 of General Comment No. 17. The right of everyone to benefit from the protection of the moral and material interests resulting from any scientific, literary or artistic production of which he or she is the author (Article 15, Paragraph 1 (c) of the ICESCR), Committee on Economic, Social and Cultural Rights, 35th session, Geneva, 7–25 November 2005.

4 Thornberry 2002.

5 See Articles 7(3) and 7(4) of the ILO Convention No. 169.

6 The Constitutive Act of the AU (AU) is available online at http://www.au2002.gov.za/docs/key_oau/au_act.htm (accessed 28 April 2009).

7 See Genugten 2008.

8 Genugten 2008.

9 This advisory opinion was quoted by Genugten 2008.

10 Genugten 2008.

11 Genugten 2008.

12 Genugten 2008.

13 Details on this process are available online at http://www.unhchr.ch/html/menu6/apw.htm (accessed 1 December 2008).

14 See http://www.unhchr.ch/html/menu6/apw.htm.(accessed 1 December 2008).

15 See http://www.unhchr.ch/html/menu6/apw.htm.

16 Available online at http://www.unhchr.ch/huricane/huricane.nsf/(Symbol)/OHCHR.STM.98.01.En?OpenDocument (accessed 1 December 2008).

17 Available online at http://www.unhchr.ch/html/menu6/islamabad.htm (accessed 1 December 2008).

18 Available online at http://india.gov.in/govt/constitutions_india_bak.php#eng (accessed 1 December 2008).

19 Article 29(1) of the Constitution of India on the Protection of the Interests of Minorities provides that 'Any section of the Indian citizens residing in the territory of India or any part thereof having a distinct language, script or culture of its own shall have the right to conserve the same.'

20 Article 244, Schedule 5 of the Constitution of India provides for the administration and control of scheduled areas and scheduled tribes (STs). Article 244(1)(4) provides for the establishment of a Tribes Advisory Council in each state having scheduled areas if the president so direct or in any state having STs but not scheduled areas therein. The advisory council shall be composed of no more than 20 members of whom three-quarter

could be members of the STs. Article 244.1.2 provides that the duty of the advisory council is to advise on such issues pertaining to the welfare and advancement of STs in the state as may be referred to them by the governor. If this type of traditional advisory council is established in many remote areas rich in traditional medicinal knowledge, they can indeed advise national authorities on how access to and use of the medicinal plants to which their knowledge is associated should be carried out. These advisory councils can therefore discuss terms of benefit sharing with those accessing their assets, perhaps with the assistance of authorities and experts knowledgeable in the commercial/market angle of the exploitation of their assets.

21 Section 3 of the Forest Rights Act extends to other types of rights that are not mentioned here. Those highlighted in this book are those that the author has found particularly relevant for the purposes of the work.

22 Article 2(g) of the Indian Forest Rights Act defines the Gram Sabha as a village assembly which shall consist of all adult members of a village and in case of states having no *panchayats, padas, tolas* or other traditional village institutions and elected village committees, with full unrestricted participation of women.

23 Article 16(a) of the Cartagena Agreement provides that the responsibilities of the Andean Council of Foreign ministers are to formulate member countries' foreign policy on matters of sub-regional interest, as well as to orient and coordinate the external efforts of the different bodies and institutions of the Andean integration system.

24 Articles 21 and 22 of the Cartagena Agreement make provisions on the composition of the Andean Community Commission and its responsibilities within the framework of the Andean Integration system.

25 See Article 129 of the Cartagena Agreement.

26 See 2001 Machu Picchu Declaration 524.

27 The 2001 Machu Picchu Declaration on Democracy, the rights of Indigenous Peoples and the War against Poverty, Andean Community (CAN), Lima, 28–9 July 2001.

28 Article 3(a) of the Decision 524 of the Andean Community.

29 Article 3(b) of Decision 524 of the Andean Community.

30 Article 8 Decision 674 of the Andean Community.

31 Article 2 Decision 674 of the Andean Community.

32 There are three generations of human rights. The first-generation human rights deal essentially with liberty and participation in political life. The human rights of the first generation are fundamentally civil and political in nature and serve to protect the individual from the excesses of the state. The second-generation human rights are related to equality and they began to be recognized after the Second World War. This generation of human rights is fundamentally composed of social, economic and cultural rights. The third-generation human rights are the rights that go beyond the mere civil and social rights as expressed in many progressive documents such as the 1972 Stockholm Declaration of the United Nations Conference on the Human Environment and the 1992 Rio Declaration on Environment and Development. Information accessed online at: http://en.wikipedia.org/wiki/Three_generations_of_human_rights (accessed 29 April 2009).

33 Section III of the Andean Community Charter for the Promotion and Protection of Human Rights of 26 July 2002. Available online at http://www.unhcr.org/refworld/docid/3de4f94a4.html (accessed 29 April 2009).

34 Section IV of the Andean Community Charter for the Promotion and Protection of Human Rights of 26 July 2002.

35 Section V of the Andean Community Charter for the Promotion and Protection of Human Rights of 26 July 2002.

36 Section VI of the Andean Community Charter for the Promotion and Protection of Human Rights of 26 July 2002.

37 Section VIII of the Andean Community Charter for the Promotion and Protection of Human Rights of 26 July 2002.

38 Article 36 of the Andean Community Charter for the Promotion and Protection of Human Rights of 26 July 2002.
39 Article 38 Andean Community Charter for the Promotion and Protection of Human Rights of 26 July 2002.
40 Article 38 Andean Community Community Charter for the Promotion and Protection of Human Rights of 26 July 2002.
41 Article 39 Andean Community Charter for the Promotion and Protection of Human Rights of 2002.
42 See Preamble of the 1967 Protocol of Buenos Aires amending the 1948 Charter which established the Organization of American States (OAS). Available online at http://www.oas.org/juridico/English/charter.html (accessed 29 April 2009).
43 Article XXVIII of the 1948 American Declaration on the Rights and Duties of Man provides that: 'the rights of man are limited by the rights of others, by the security of all, and by the just demands of the general welfare and the advancement of democracy'. Available online at http://www.hrcr.org/docs/OAS_Declaration/oasrights4.html (accessed 30 April 2009).
44 Thornberry 2002.
45 An online version of the Inter American Convention on Human Rights is available at: http://www.hrcr.org/docs/American_Convention/oashr5.html (last accessed 30 April 2009).
46 Thornberry 2002.
47 Thornberry 2002.
48 Thornberry 2002.
49 Draft Resolution: Draft American Declaration on the Rights of Indigenous Peoples. Permanent Council of the OAS. Committee on Juridical and Political Affairs, OEA/Ser. G, CP/CAJP-2638/08, 14 May 2008.
50 Draft Resolution: Draft American Declaration on the Rights of Indigenous Peoples, OEA/Ser.G, CP/CAJP-2638/08, 14 May 2008.
51 Draft Resolution: Draft American Declaration on the Rights of Indigenous Peoples, OEA/Ser.G, CP/CAJP-2638/08, 14 May 2008.
52 There are indeed references to the collective nature of the rights or collective nature of the systems of indigenous peoples pertaining to their mode of utilization of their assets in the 2000 version of the OAS Proposed Declaration on the Rights of Indigenous Peoples. For instance, paragraph 5 of the introduction to the proposed declaration stresses its recognition of the traditional collective systems of indigenous culture for the use and control of their land and territories and collective forms of land ownership and social organizations. See Proposed American Declaration on the Rights of Indigenous Populations, Considered at the Meetings Held on 8–12 November 1999 Incorporating the Observations and Proposals of Representatives of Indigenous Populations' Permanent Council of the OAS, Committee on Juridical and Political Affairs, Working Group to Prepare the Proposed American Declaration on the Rights of Indigenous Populations, OEA/Ser.K/XVI, GT/DADIN/doc.1/99 rev. 2, 6 December 2000.
53 Thornberry 2002.
54 For an insight into the various debates that surrounded the development of this instrument, see Asbjorn 2007.
55 See Doc. Assembly/AU/Dec.141 (VIII), 30 January 2007.
56 UN Docs. E/CN.4/832/Rev.1, Paragraph 105. This paragraph is based on the 1962 statement of the United Nations legal adviser.
57 The Declaration on the Rights of Indigenous Peoples, United Nations, New York, 2007.
58 Article 18 of the 2007 UN-DRIP.
59 Article 7 of the 2007 UN Declaration on the Rights of Indigenous Peoples.
60 Article 5 of the 2007 UN Declaration on the Rights of Indigenous Peoples.
61 Article 14 of the 2007 UN Declaration on the Rights of Indigenous Peoples.
62 Article 24 of the 2007 UN Declaration on the Rights of Indigenous Peoples.
63 Article 31 of the 2007 UN Declaration on the Rights of Indigenous Peoples.

64 Articles 32, 36 and 38 of the 2007 UN Declaration on the Rights of Indigenous Peoples.
65 Articles 10, 11, 19, 28, 29, 32 of the 2007 UN Declaration on the Rights of Indigenous Peoples.

5 Incursion in the 'bio-piracy' debate

1 Dutfield 2008.
2 Dreher *et al.* 2000.
3 Dreher *et al.* 2000.
4 Dutfield 2008: 29.
5 Dutfield 2008:29.
6 Dreher *et al.* 'Is Marker-Assisted Selection Cost-Effective?', p. 2.
7 Halewood and Nnadozie 2008.
8 Dickson and Gagnon 2004.
9 Dickson and Gagnon 2004.
10 Laird and ten Kate 2002.
11 Cragg *et al.* 1997.
12 Newman *et al.* 2003
13 See PhRMA 2003.
14 Arnold 2007.
15 Davies-Coleman and Beukes 2004.
16 Email communication with a South African expert (whose anonymity I want to preserve) on 17 May 2007.
17 Laird and ten Kate 2002.
18 Paramasivam 2008.
19 See http://www.hashmi.com/turmeric.html (accessed 28 December 2008).
20 Mgbeoji 2006.
21 Schuler 2004.
22 See http://www.healthydirect.com/product/TURMC/d.jsf;jsessionid=8E6611BA35E1EE8CD0500C7C4EACAD05.www1?AffiliateCode=GooglePPC&_s_ref=ZhOn6vL6S&kw=turmeric&creative=1467249482&gclid=CNeD3PC345cCFUog3godBlntCw (accessed 28 December 2008).
23 See http://newsletter.vitalchoice.com/e_article000689711.cfm?x=b8mTBjS,b61qBrB1 (accessed 28 December 2008).
24 See http://www.lancospices.com/ (accessed 28 December 2008).
25 I acknowledge that there is no indication that any corporation doing business on turmeric products or that any turmeric-related patent has tried to stop Indian's traditional communities from using the roots of turmeric plants in their own way or selling their traditionally processed products.
26 Koyama and Mayet 2007.
27 Koyama and Mayet 2007.
28 Koyama and Mayet 2007.
29 Koyama and Mayet 2007.
30 Koyama and Mayet 2007.
31 See Patent WO/2008/125239, available online at http://www.wipo.int/pctdb/en/ia.jsp?ia=EP2008/002720 (accessed online 13 May 2009).
32 Koyama and Mayet 2007. The other countries are Australia, Brazil, Canada, China, Japan, Korea, Mexico, New Zealand, Philippines, Russia, and the USA.
33 See, for instance, US National Research Council (1989).
34 See http://www.macaroot.com/history/articles.html (accessed 5 January 2009).
35 See http://www.macaroot.com/history/articles.html.
36 See http://www.naturex.com/actionnaires/gb/pressefi_gb.htm?page==http://www.actusnews.com/export_communique.php?CLIENT=ACTUS-0-71&ID=ACTUS-0–5098&logo=1&tab=102 (accessed 6 January 2009).

37 See http://www.naturex.com/actionnaires/gb/pressefi_gb.htm?page==http://www.actusnews.com/export_communique.php?CLIENT=ACTUS-0-71&ID=ACTUS-0–5098&logo=1&tab=102.
38 See http://www.naturex.com/actionnaires/gb/pressefi_gb.htm?page==http://www.actusnews.com/export_communique.php?CLIENT=ACTUS-0-71&ID=ACTUS-0-5886&logo=1&tab=102 (accessed 6 January 2009).
39 See http://www.naturex.com/actionnaires/gb/pressefi_gb.htm?page==http://www.actusnews.com/export_communique.php?CLIENT=ACTUS-0-71&ID=ACTUS-0–13911&logo=1&tab=102 (accessed 6 January 2009).
40 See http://www.naturex.com/actionnaires/gb/pressefi_gb.htm?page==http://www.actusnews.com/export_communique.php?CLIENT=ACTUS-0-71&ID=ACTUS-0-13479&logo=1&tab=102 (accessed 6 January 2009).
41 See http://www.mindfully.org/GE/GE4/Maca-Patents-Peruvian-Farmers3jul02.htm (accessed 6 January 2009).
42 See http://www.mindfully.org/GE/GE4/Maca-Patents-Peruvian-Farmers3jul02.htm.
43 See http://www.wipo.int/pctdb/cgi/guest/search5 (accessed 6 January 2009).
44 See http://www.wipo.int/pctdb/en/fetch.jsp?LANG=ENG&DBSELECT=PCT&SERVER_TYPE=19–00&SORT=41252552-KEY&TYPE_FIELD=256&IDB=0&IDOC=1096171&C=00&ELEMENT_SET=BASICHTML-ENG&RESULT=12&TOTAL=12&START=1&DISP=25&FORM=SEP-0/HITNUM,B-ENG,DP,MC,AN,PA,ABSUM-ENG&SEARCH_IA=US2000005607&QUERY=%22Lepidium%22%0d%0a (accessed 6 January 2009).
45 See http://www.wipo.int/pctdb/en/fetch.jsp?SEARCH_IA=IB2007002046&DBSELECT=PCT&C=00&TOTAL=12&IDB=0&TYPE_FIELD=256&SERVER_TYPE=19–00&QUERY=%22Lepidium%22%0D%0A&ELEMENT_SET=BASICHTML-ENG&START=1&SORT=41252552-KEY&RESULT=3&DISP=25&FORM=SEP-0%2FHITNUM%2CB-ENG%2CDP%2CMC%2CAN%2CPA%2CABSUM-ENG&IDOC=562101&IA=IB2007002046&LANG=ENG&DISPLAY=DESC (accessed 6 January 2009).
46 See http://www.patentstorm.us/patents/6267995.html (accessed 6 January 2009).
47 See http://www.patentstorm.us/patents/6267995.html.
48 See http://www.mindfully.org/GE/GE4/Maca-Patents-Peruvian-Farmers3jul02.htm (accessed 6 January 2009).
49 See http://www.mindfully.org/GE/GE4/Maca-Patents-Peruvian-Farmers3jul02.htm.
50 Balee 1994.
51 It has been reported that only in the first quarter of 2002, more than 50 tons of Cupuaçu seeds were imported by Japan to fuel the 'cupulate' industry and more than 200 tons were to be imported by Japan from Brazil in 2003. See www.amazonlink.org/biopiracy/cupuacu.htm (accessed 2 February 2009).
52 See http://www.amazonlink.org/biopiracy/cupuacu.htm (accessed 9 January 2009).
53 Reisdorff, *et al.* 2000.
54 See various patents at http://www.google.com/patents?q=Cupuacu+patents&scoring=1&output=rss (accessed 9 January 2009).
55 See US patent on Tropical Compositions Comprising Myrica Gail oil at http://appft1.uspto.gov/netacgi/nph-Parser?Sect1=PTO2&Sect2=HITOFF&p=1&u=%2Fnetahtml%2FPTO%2Fsearch-bool.html&r=1&f=G&l=50&co1=AND&d=PG01&s1=Cupuacu&s2=%22theobroma+grandiflorum%22&OS=Cupuacu+AND+"theobroma+grandiflorum"&RS=Cupuacu+AND+"theobroma+grandiflorum" (accessed 9 January 2009).
56 See Munoz Tellez 2006.
57 Munoz Tellez 2006.
58 See Schmidlehner 2003.

6 Soft and regional undertakings aimed at community rights

1 Dutfield 2002a.
2 Dutfield 2002a.

3 Centre for International Environmental Law 2000.
4 See Feldbaum 2002.
5 International Chamber of Commerce 2001.
6 International Chamber of Commerce 2001.
7 A synthesis of the CUSCO Declaration is found in International Centre for Trade and sustainable Development 2002.
8 The countries include Bolivia, Brazil, China, Colombia, Costa Rica, Ecuador, the Philippines, India, Indonesia, Kenya, Malaysia, Mexico, Peru, South Africa and Venezuela. Available online at http://www.undp.org/biodiversity/events/Megadiverse_Meeting.html (accessed 26 January 2009).
9 Latorre *et al.* 2001.
10 Laird and ten Kate 2002.
11 The US-based Business for the Social Responsibility (BSR) is a network aiming to help business companies to be 'commercially successful in ways that demonstrate respect to ethical values, people and communities and the environment'. Companies involved in the BSR undertake a range of activities in the global economy. See http://www.bsr.org.
12 Halewood 1999; Timmermans 2003.
13 Halewood 1999.
14 Article 28(3) of ILO 169 stipulates that measures shall be taken to promote the development and practice of indigenous languages of the people concerned.
15 Article 8(j) of the 1992 Rio CBD.
16 Posey 1996.
17 Posey 1995.
18 A Canadian indigenous peoples organization, with consultative status to the United Nations Economic and Social Council. See http://www.unhchr.ch/html/racism/indileaflet1.doc (accessed 14 May 2009). This definition was submitted to the Secretariat of the CBD in 1996.
19 Point 4 of the Declaration of Belem adopted at the First International Congress of Ethnobiology in Belem, Brazil, in 1988 calls for the development of procedures to compensate native peoples for the utilization of their knowledge and their biological resources. The declaration of Belem is available at http://ise.arts.ubc.ca/global_coalition/declaration.php (accessed 14 May 2009).
20 The International Society of Ethnobiology Code of Ethics that was revised in New Zealand in November 1998. The code of ethics is a set of principles including the principle of prior rights, principle of self-determination, principles of traditional guardianship, principle of compensation and equity in benefit sharing. The Code of Ethics based on the Declaration of Belem, is available at http://ise.arts.ubc.ca/_common/docs/ISE%20COE_Eng_rev_24Nov08.pdf (accessed 14 May 2009).
21 Greene 2002.
22 Greene 2002.
23 Ekpere 2001.
24 Ekpere 2001.
25 Ekpere, 2001 Article 18.
26 Ekpere, 2001 Article 19.
27 Ekpere, 2001 Article 20.
28 Ekpere, 2001 Articles 21 and 22.
29 Ekpere, 2001 Article 8.
30 Ekpere, 2001 Article 9.
31 Preamble of Decision 391 on the Common Regime on Access to Genetic Resources of the Andean Community.
32 Articles 16 and 17 of the Common Regime on Access to Genetic Resources of the Andean Community.
33 Article 1 of Decision 391 on the Common Regime on Access to Genetic Resources of the Andean Community.

34 Article 2(a) of the Andean Common Regime on Access.
35 Article 2(b) of the Andean Common Regime on Access.
36 Article 34 of the Andean Common Regime on Access.
37 Article 45 of the Andean Common Regime on Access.

7 Broader framework of the suggested regulatory measures

1 Halewood 1999; Timmermans 2003.
2 The preamble of the 1981 African Charter addresses equality in relation to freedom, justice and dignity among the essential objectives of the Charter. More specifically, Article 13 of the Charter stipulates that all peoples must be able to access all public property and services on an equal basis before the law.
3 Article 26 of the 1996 United Nations International Covenant on Civil and Political Rights (UN ICCPR) merges equality and non-discrimination in the following terms: 'all persons are equal before the law and are entitled without any discrimination to equal protection of the law. In this respect, the law shall prohibit any discrimination and guarantee to all persons equal and effective protection against discrimination on any ground.'
4 The 1989 International Labour Organization (ILO) Convention 169 Concerning Indigenous and Tribal Peoples in Independent Countries has specific provisions supportive to national regulatory measures addressing full participation of rural communities in decision-making processes relating to issues impacting their lives. See Articles 6(a) and Article 6(2) of the 1989 ILO Convention 169.
5 Articles 21(1) and 21(5) of the 1981 African Charter can be interpreted to suggest that the communities' rights to benefit from the exploitation of national biological resources must be recognized as human rights.
6 The 1981 African Charter addresses self-determination in Article 20. Article 1(2) of ILO Convention 169 addresses 'Self-identification', which can be interpreted as another variant of 'Self-determination'. Article 3 of the 2007 UN Declaration addresses self-determination in the sense of ensuring empowerment through the promotion of self-governance at the local level.
7 Heald 2005.
8 Articles 15(5) and 8(j) of the Convention on Biological Diversity (CBD) and Articles 24–40 of the Bonn Guidelines. Although it refers to the state as the contracting party that may provide consent for access to genetic resources, Article 8(j) implies that despite being subject to national legislation, local communities, who are the holders of traditional knowledge (TK), innovations and practices, should be involved in and even approve the decisions granting access to the resources under their custody.
9 Erosion Technology and Concentration 2004. There are criticisms about the efficiency of these instruments in ensuring that the prior informed consent (PIC) of local communities on issues affecting them is actually sought. In one communiqué, the action group on Erosion, Technology and Conservation (ETC) asserts that the CBD, the Bonn guidelines and even national legislation on access and benefit sharing (ABS) are basically tools that have strengthened the enclosure of the 'commons' that started during colonization.
10 Correa 2004.
11 See Chapter 3 on marked failures of the Brazilian and Peruvian biodiversity measures to mention that acquiring community assets through means that have failed to incorporate the PIC of local communities are punishable offences under the laws.
12 Annex 1, International Regime, Section C on Compliance of Decision IX/12 Access and Benefit-Sharing. UNEP/CBD/COP9/9/29.
13 See CITES Compliance Procedures adopted at the 14th Conference of the Parties on 3–15 June 2007 in The Hague. Available online at http://www.cites.org/eng/res/14/14–03.shtml (accessed 26 January 2009).

14 CITES Compliance Procedures, Article 30.
15 Dinwoodie 2004.
16 Article 6(1), Convention for the Protection of Industrial Property (Paris, 20 March 1883).
17 Convention for the Protection of Industrial Property, Article 6 quinquies, provides that 'every trademark duly registered in the country of origin shall be accepted for the filing and protected as is in other countries of the Union subject to the reservations indicated in this article. Such countries may, before proceeding to final registration, require the production of a certificate of registration in the country of origin issued by the competent authority. No authentication shall be required for this certificate.'
18 Convention for the Protection of Industrial Property Article 6 quinquies (B)(1). This article allows the states to refuse recognition of foreign trademark rights in certain circumstances.
19 Austin 2001.
20 Austin 20001.
21 According to Neil Andrews, civil cases are subject to the mechanics of international private law including the terms of choice of forum and choice of applicable law. See Andrews 1994.
22 World Intellectual Property Organization 2001a.
23 Dogauchi 2001.
24 Dogauchi, 2001.
25 Savigny 1849.
26 Giliker and Beckwith, 2004.
27 Giliker and Beckwith 2004.
28 Supreme Court Judgment on 16 October 1981, *Michiko Goto, et al.* v. *Malaysian Airlines* 'System Berhadin', 35(7) *Minshu,* 1224.
29 For the jurisdictional grounds under the Japanese CCPs Law No. 109, 1996, see Chapter 2 (Court), Section 1 (Jurisdiction) Articles 4 and 5 of the revised CCP of Japan promulgated in 1996. Available online at http://www.japanlaw.co.jp/procedure/civilcode.html (accessed 28 January 2009).

8 Applicability of the regulatory measures

1 Food and Agriculture Organization of the United Nations 1981.
2 'The Rio Declaration on Environment and Development', the United Nations Conference on Environment and Development, 3–14 June 1992, Rio de Janeiro, Brazil.
3 The 1992 United Nations Rio Declaration on Environment and Development.
4 Regulations on Bioprospecting, Access and Benefit-Sharing, Government Gazette, No. 30739, 8 February 2008.
5 Tonye 2005.
6 Regulations 6, Chapter 2 of the South African Biodiversity Regulations 2008.
7 Chapter 8, Part 1, Section 97(b)(i) of the South African Biodiversity Act No. 10 of 2004.
8 Information gathered during an interview with Sarah Laird in October 2004 in London, and with Roger Chennells in November 2004 in Cape Town.
9 This statement was made by Sarah A. Laird during discussion with her in London in October 2004. Sarah A. Laird is an international consultant on biodiversity policy-making.
10 See http://www.biowatch.org.za (accessed 15 April 2009).
11 See http://www.aiccafrica.com (accessed 15 April 2009).
12 National Environmental Management: Biodiversity Act No. 10, 2004, Chapter 7, Part 1, Section 88 (2)(c) addressing application for permits.
13 Burgener 2003.
14 For an exhaustive list of elements that characterize unpatentable inventions, see Sections 25(2), (3), (4)(a), (4)(b), (11) of the Patent Act 57 of 1978.
15 Section 25(4)(b) of the South African Patent Act 57 of 1978.

16 Burrell 1972.
17 J. Younger (1916).
18 Burrell 1972.
19 The Patent Amendment Act 20, 2005, *Government Gazette,* Vol. 486, No. 28319, South Africa.
20 The South African Patent Amendment Act 20, 2005, Section 3(B).
21 Industrial Property Law, 14 May 1996, No. 9(279). Article 8.
22 By 'biodiversity laws' here I refer to the South African Biodiversity Act No. 10 of 2004 and the Brazilian Provisional Measure on 'Access to Genetic Resources and Traditional Knowledge', No. 2186–16, 2001. The Brazilian Provisional measures were discussed in Chapter 3.
23 Lewis and Ramani 2003; also, the Access and Benefit Sharing as Related to Genetic Resources (Decision COP VI/24, 19 April 2002), Annex (Bonn Guidelines on Access to Genetic Resources and Fair and Equitable Sharing of the Benefits Arising out of their Utilization).
24 Visser 2004.
25 The production and sales of semi-synthetic versions is indeed a form of direct use and potential direct benefits from the patent on P57 that was licensed to the company by the South African Council for Scientific and Industrial Research (CSIR). The revenues of Phytopharm related to the sales of the semi-synthetic products from the P57 may not be known to the San communities neither to CSIR as the company is free to establish partnerships with potential interested parties. See the press release of Phytopharm, 30 July 2002, available online at http://www.phytopharm.com/press/New%20arrangements%20final.htm (accessed December 2005).
26 There are growing twists and turns in the development process of functional food targeting weight management based on hoodia products. In December 2008, Unilever decided to return the rights over hoodia extracts to Phytopharm, meaning that the company is withdrawing from the development of the weight-management functional food based on hoodia. See press release, December 2008, Unilever Returns rights on hoodia, extract available online at http://www.phytopharm.com/news/newsreleases/?id=16553 (accessed 19 May 2009). These twists and turns mean that the San communities are highly likely to wait for longer to see the completion and commercialization of hoodia-based products upon which they will earn royalties from Phytopharm.
27 Benefit Sharing Agreement Between the CSIR (A Statutory Council Established in Terms of Act No. 46 of 1988 herein Represented by Dr. Sibusiso Sibisi in his Capacity as President of the CSIR) and the South African San Council A Voluntary Association Established by = Khomani, !Xun and the Khwe San Communities of South Africa in Terms of a Constitution Signed and Adopted on the 27 November 2001, at Kimberley and herein Represented by Petrus Vaalbooi in his Capacity as Chairperson and Duly Authorized Thereto.
28 Kono 2000.
29 Article 10(3) of The Hague Convention 2005 that makes reference to Article 2(2), excludes tort and intellectual property rights other than copyrights and related rights from its coverage. The Hague Convention on Choice of Court Agreement, which is not yet in force (The Hague, June 2005), available online at http://www.hcch.net/index_en.php?act=conventions.text&cid=98 (accessed 16 February 2009).
30 Drahos 2004.
31 Braithwaite and Drahos 2000. In this book, the authors address the use of Memorandum of Understanding (MoU) in other areas of international regulations such as drugs, food and competition law.
32 Drahos 2004.
33 The Convention on the Recognition and Enforcement of Foreign Arbitral Awards (New York, 1958).

34 Important aspects of litigation processes under the supervision of the centre include the choice of the venue of litigation as well as the choice of applicable law. Both the venue and the applicable law may not be related to Cameroon or Japan. A neutral country whose laws will be applicable laws can be chosen. See http://www://arbiter.wipo.int (accessed 14 February 2009).

35 See http://www.unidroit.org/english/principles/contracts/main.htm (accessed 16 February 2009).

36 The term agreement refers to 'agreement in writing' as defined under Article II of the 1958 New York Convention.

Bibliography

Andersen, R. (2007) *Protecting Farmers' Rights in the Global IPR Regime: Challenges and Options for Developing Countries,* Policy Brief for the South Asia Watch on Trade, Economics & Environment, Kathmandu: SAWTEE.

Andersen, R. and Winge, T. (2008) *Background Study 7: Success Stories from the Realization of Farmers' Rights Related to Plant Genetic Resources for Food and Agriculture,* Lysaker, Norway, Fridtjof Nansens Institute/The Development Fund.

Andrews, N. (1994) *Principles of Civil Procedure,* London: Sweet & Maxwell.

Arnold, R. J. G. (2007) 'Cost-Effectiveness Analysis: Should It Be Required for Drug Registration and Beyond?' *Drug Discovery Today,* 12 (21–2): 960.

Asbjorn, E. (2007) 'Rights of Indigenous Peoples: Achievements in International Law during the Last Quarter of a Century', *Netherlands Yearbook of International Law, 2006,* Vol. XXXVII, The Hague: TMC Asser Press, pp. 155–212.

Austin, G. (2001) *Private International Law and Intellectual Property Rights: A Common Law Overview,* WIPO Forum on Private International Law and Intellectual Property, Geneva, 30–1 January, WIPO/PIL/01/5.

Bagley, M. A. (2003) 'Patently Unconstitutional: The Geographical Limitation on Prior Art in a Small World', *Minnesota Law Review,* 87: 679–727.

Balee, W. (1994) *Footprint of the Forest,* New York: Columbia University.

Bell, R. H. V. (1987) 'Conservation with a Human Face: Conflict and Reconciliation in African Land Use Planning', in D. Anderson and R. Grove (eds), *Conservation in Africa: People and Practice,* Cambridge: Cambridge University Press, pp. 79–102.

Bengwayan, M. A. (2003) *Intellectual and Cultural Property Rights of Indigenous and Tribal Peoples in Asia,* London: Minority Rights Group International (MRGI).

Bently, L. and Sherman, B. (2001) *Intellectual Property Law,* Oxford: Oxford University Press.

Bessen, J. (2005) 'Patents and the Diffusion of Technical Information', *Economic Letters,* 86 (1): 121–8.

Blakeney, M. (1999) 'Ethnobiological Knowledge and the Intellectual Property Rights of Indigenous Peoples in Australia', in M. Blakeney (ed.), *Intellectual Property Aspects of Ethnobiology,* London: Sweet & Maxwell, pp. 83–100.

Boyko, R. D. (1998) 'Antitrust Limits on Exploiting Intellectual Property Rights', *St John's Journal of Legal Commentary,* 13 (2): 427–56.

Braithwaite J. and Drahos, P. (2000) *Global Business Regulation,* Cambridge: Cambridge University Press.

Brokensha, D. and Riley, B. W. (1980) 'Mbeere Knowledge of their Vegetation and its Relevance for Development: A Case Study from Kenya', in D. Brokensha, D. M.

Warren and O. Werner (eds), *Indigenous Knowledge and Development,* Lanham, Md.: University Press of America, pp. 111–28.

Brown, M. F. (2003) *Who Owns Native Culture,* Cambridge, Mass.: Harvard University Press.

Brush, S. B. (1993) 'Indigenous Knowledge of Biological Resources and Intellectual Property Rights: The Role of Anthropology', *American Anthropologist,* 95 (3): 653–86.

Burgener, M. (2003) 'A Review of the Existing South African Administrative Systems for Permitting and Overview on Benefit Sharing Schemes', in IUCN (ed.), *Developing Access and Benefit Sharing Legislation in South Africa: A Review of International and National Experiences,* Gland: IUCN, pp. 82–102.

Burrell, T. D. (1972) *South African Patent Law and Practice,* Durban: Butterworths South Africa.

Cantuaria Marin, P. L. (2002) *Providing Protection for Plant Genetic Resources: Patents, Sui Generis Systems and Biopartnerships,* New York: Kluwer Law International.

Centre for International Environmental Law (2002) *Community Based Property Rights: A Concept Note, Centre for International Environmental Law,* Washington, DC: Centre for International Environmental Law.

Clarke, J., Cavendish, W. and Coote, C. (1996) 'Rural Households and Miombo Woodlands: Use, Value and Management', in B. Campbell (ed.), *The Miombo in Transition: Woodlands and Welfare in Africa,* Bogor, Indonesia: Centre for International Forestry Research, pp. 101–35.

Coase, R. H. (1960) 'The Problem of Social Cost', *Law & Economics,* 3 (1): 11.

Correa, C. (2004) 'Protecting Traditional Knowledge: Lessons from National Experiences', prepared for the UNCTAD Meeting, Geneva, 4–6 February 2004.

Cragg, G. M., Newman, D. J. and Sander K. M. (1997) 'Natural Products in Drug Discovery and Development', *Journal of Natural Products,* 6 (1): 52–60.

Cullet, P. (2005) *Intellectual Property Protection and Sustainable Development,* New Delhi: LexisNexis Butterworths.

Davies-Coleman, M. T. and Beukes, D. R. (2004) 'Ten Years of Marine Natural Products Research at Rhodes University', *South African Journal of Science,* 100: 539–44.

Demsetz, H. (1967) 'Towards a Theory of Property Rights', *American Economic Review,* 57 (2): 347–59.

Dhar, B. (2002) *Sui Generis Systems for Plant Variety Protection, Options under TRIPS: A Discussion Paper,* Geneva: Quaker United Nations Office.

Dickson, M. and Gagnon, J. P. (2004) 'The Cost of New Drug Discovery and Development', *Discovery Medicine,* 4 (22): 172–9.

Dinwoodie, G. B. (2004) 'Towards an International Framework for the Protection of Traditional Knowledge', paper submitted at the UNCTAD Meeting, Geneva, 4–6 February 2004.

Dogauchi, M. (2001) 'Private International Law on Intellectual Property: A Civil Law Overview', WIPO Forum on Private International Law and Intellectual Property, Geneva, 30–1 January.

Drahos, P. (2004) 'Towards an International Framework for the Protection of Traditional Group Knowledge and Practice', paper submitted at the UNCTAD Meeting, Geneva 4–6 February 2004.

Dreher, K., Morris, M., Khairallah, M., Ribaut, J.-M., Pandey, S. and Srinivasam, G. (2000) *Is Marker-Assisted Selection Cost-Effective Compared to Conventional Breeding Methods? The Case of Quality Protein Maize,* Mexico City: International Maize and Wheat Improvement Center.

Dutfield, G. (1999) 'Protecting and Revitalizing Traditional Ecological Knowledge: Intellectual Property Rights and Community Knowledge in India', in M. Blakeney (ed.), *Intellectual Property Aspects of Ethnobiology,* London: Sweet & Maxwell, pp. 101–22.

——(2002a) 'Indigenous People's Declarations and Statements and Equitable Research Relationships', in S. A. Laird (ed.), *Biodiversity and Traditional Knowledge: Equitable Partnerships in Practice,* London: Earthscan, pp. 228–32.

——(2002b) *Protecting Traditional Knowledge and Folklore: A Review of Progress in Diplomacy and Policy Information,* ICTSD-UNCTAD Project on IPRs.

——(2002c) 'Sharing the Benefits of Biodiversity: Is There a Role for the Patent System?' *Journal of World Intellectual Property,* 5 (6): 899–931.

——(2004) *Intellectual Property, Biogenetic Resources and Traditional Knowledge,* London: Earthscan.

——(2005) *Thinking Aloud on Disclosure of Origin,* Occasional Paper 18, Geneva: Quaker United Nations.

——(2008) 'Turning Plant Varieties into Intellectual Property: The UPOV Convention', in G. Tansey and T. Rajotte (eds), *The Future Control of Food: A Guide to International Negotiations and Rules on Intellectual Property, Biodiversity and Food Security,* London: Earthscan, pp. 27–47.

Ekpere, J. A. (2001) *The African Model Law: The Protection of the Rights of Local Communities, Farmers and Breeders, and for the Regulation of Access to Biological Resources: An Explanatory Booklet,* Organization of African Unity.

Erosion Technology and Concentration (2004) 'From Global Enclosure to Self-Enclosure: Ten Years After – A Critique of the Bonn Guidelines on Access and Benefit Sharing (ABS)', ETC group communiqué, 83 (January/February).

European Chemical Industry Council (2002) *CEFIC Position Paper: The Chemical Industry Comments on the Legal Protection of Traditional Knowledge and Access to Genetic Resources-Patenting,* Brussels: European Chemical Industry Council.

Farnsworth, N. R., Akerele, O., Bingel, A. S., Soejarto, D. D. and Guo, Z. (1985) 'Medicinal Plants in Therapy', *Bulletin of the World Health Organization,* 63 (6): 965–81.

Feldbaum, C. B. (2002) 'Biotechnology Foreign Policy', available online at http://www.bio.org/speeches/speeches/20020610.asp (accessed 14 May 2009).

Field-Juma, A. (1996) 'Governance and Sustainable Development', in C. Juma and J. B. Ojwang (eds), *In Land We Trust: Environment, Private Property and Constitutional Change,* Nairobi and London: Initiatives Publishers and Zed Books, pp. 117–42.

Food and Agriculture Organization of the United Nations (1981) *The Peasant's Charter: The Declaration of Principles and Programme of Action of the World Conference on Agrarian Reform and Rural Development,* Rome: Food and Agriculture Organization of the United Nations.

Genugten, W. van (2008) 'The African Move Towards the Adoption of the 2007 Declaration on the Rights of Indigenous Peoples: The Substantive Arguments behind the Procedures', paper prepared for the Committee on the Rights of Indigenous Peoples of the International Law Association, of which of the author is a member, March. Available online at http://papers.ssrn.com/sol3/papers.cfm?abstract_id=1103862 (accessed 28 April 2009).

Geroski, P. (1995) 'Markets for Technology: Knowledge, Innovation and Appropriability', in P. Stoneman (ed.), *Handbook of the Economics of Innovations and Technological Change,* Oxford and Malden, Mass.: Blackwell, pp. 90–131.

Gibson, J. (2005) *Community Resources: Intellectual Property, International Trade and Protection of Traditional Knowledge,* London: Ashgate.

——(2008) *Indigenous Peoples Boycott at WIPO,* posted on IKPAT blog on Friday, 17 October, available online at http://ipkitten.blogspot.com/2008/10/indigenous-peoples-boycott-at-wipo.html (accessed 9 April 2009)

Giliker, P. and Beckwith, S. (2004) *Tort,* London: Sweet & Maxwell.

Gold, R. E. and Lam, K. D. (2003) 'Balancing Trade in Patents: Public Non-Commercial Use and Compulsory Licence', *Journal of World Intellectual Property,* 6 (1): 5–32.

Greene, S. (2002) 'Intellectual Property, Resources or Territory? Reframing the Debate Over Indigenous Rights, Traditional Knowledge, and Pharmaceutical Bioprospection', in P. B. Mark and P. Patrice (eds), *Truth Claims: Representations and Human Rights,* New Brunswick, NJ: Rutgers University Press, pp. 229–49.

Griffiths, T. (1993) 'Indigenous Knowledge and Intellectual Property: A Preliminary Review of Anthropological Literature', unpublished report.

Gupta, A. (2002) 'Is All TK a Prior Art? How to Make IPR Regime Responsive to the Needs of Small, Scattered and Disadvantaged Innovators and Traditional Knowledge Holders: Honey Bee Experience', available online at http://www.sristi.org/anilg/files_1979_2003/Is%20all%20TK%20a%20prior%20Art.PDF (accessed 8 April 2009).

Gupta, A. K. (2004) 'WIPO-UNEP Study on the Role of Intellectual Property Rights in the Sharing of Benefits Arising from the Use of Biological Resources and Associated Traditional Knowledge', available online at http://www.wipo.int/tk/en/publications/769e_unep_tk.pdf (accessed 6 April 2009).

Halewood, M. (1999) 'Indigenous and Local Knowledge in International Law: A Preface to Sui Generis Intellectual Property Protection', *McGill Law Journal,* 44 (4): 952–96.

Halewood, M. and Nnadozie, K. (2008) 'Giving Priority to the Commons: The International Treaty on Plant Genetic Resources for Food and Agriculture', in G. Tansey and T. Rajotte (eds), *The Future Control of Food: A Guide to International Negotiations and Rules on Intellectual Property, Biodiversity and Food Security,* London: Earthscan, pp. 115–40.

Hart, T. and Fazzani, L. (2000) *Intellectual Property Law,* 2nd edn, New York: Palgrave Law Master.

Heald, P. J. (2005) 'A Transaction Costs Theory of Patent Law', *Ohio State Law Journal,* 66 (3): 473–96.

Helfer, L. R. (2002) Intellectual Property Rights in Plant Varieties: An Overview with Options for National Governments, FAO Legal Papers Online No. 31, July. Available online at http://www.fao.org/Legal/Prs-OL/lpo31.pdf (accessed 20 February 2009).

Heywood, V. H. and Watson, R. T. (1995) *Global Biodiversity Assessment,* Cambridge: Cambridge University Press.

Hladik, C. M., Hladik, A., Limares, O. F., Pagezy, H., Semple, A. and Hadley, M. (eds) (1993) *Tropical Forest, People and Food: Biocultural Interactions and Applications to Development,* Carnforth and Paris: Parthenon Publishing and UNESCO.

Hodge, J. (2000) 'Comment 1126 Claim Interpretation and the Doctrine of Equivalents: An Invitation to Confused Thinking', *Santa Clara Computer & High Technology Law Journal* 17: 203–29.

Honore, A. M. (1961) 'Ownership', in A. G. Guest (ed.), *Oxford Essays in Jurisprudence,* Oxford: Oxford University Press, pp. 107–47.

Hughes, J. D. (1998) 'Sacred Groves of the Ancient Mediterranean Area: Early Conservation of Biological Diversity', in P. S. Ramakrishnan, K. G. Saxena and U. M. Chandrashekara (eds), *Conserving the Sacred for Biodiversity Management,* New Delhi, Oxford and Paris: IBH Publishing and UNESCO, pp. 101–21.

International Centre for Trade and Sustainable Development (2002) 'Megadiverse Countries Set Up Working Group on Benefit Sharing', *Bridges Trade BioRes,* 2 (19).

International Chamber of Commerce (2001) *Protecting Traditional Knowledge: Discussion Paper Prepared by the Commission on Intellectual and Industrial Property,* Document No 450/937 Rev., International Chamber of Commerce.

Kameri-Mbote, P. (2003) 'Community, Farmers' and Breeders' Rights in Africa: Towards a Legal Framework for Sui Generis Legislation', *University of Nairobi Law Journal,* 1: 120–45.

Kloppenburg, J., Jr. and Kleinman, D. L. (1987) 'The Plant Germplasm Controversy', *BioScience,* 37 (3): 190–8.

Kono, T. (2000) 'Enforcement of Foreign Intellectual Property Rights in Japan: Jurisdiction and Applicable Law', CASRIP Publication Series Rethinking International Intellectual Property 6. Available online at http://www.law.washington.edu/casrip/Symposium/Number6/Kono.pdf (accessed 24 September 2005).

Koopman, J. (2008) 'Human Rights Implications of Patenting Biotechnological Knowledge', in P. L. C. Torremans (ed.), *Intellectual Property and Human Rights,* Alphen: Kluwer Law International, pp. 533–82.

Koyama, M.M. and Mayet, M. (2007) *Bioprospecting, Biopiracy and Indigenous Knowledge. Two case studies from the Eastern Cape Province*, South Africa, Johannesburg: African Centre for Biosafety.

Kushan, J. (2002) 'Patents and Biopiracy: Getting the Bottom of the Debate', Presentation at the Conference The International Patent System, Geneva, 26 March 2002.

Kuyek, D. (2002) 'Intellectual Property Rights in African Agriculture: Implications for Rural Farmers', Barcelona: Genetic Resources Action International.

Laird, S. A., Johnston, S., Wynberg, R., Lysinge, E. and Lohan, D. (2003) *Biodiversity Access and Benefit Sharing Policies for Protected Areas: An Introduction,* Tokyo: United Nations University Institute of Advanced Studies.

Laird, S. A. and ten Kate, K. (2002) 'Biodiversity Prospecting: The Commercial Use of Genetic Resources and Best Practice in Benefit-Sharing', in S. A. Laird (ed.), *Biodiversity and Traditional Knowledge: Equitable Partnerships in Practice,* London: Earthscan, pp. 241–85.

Lane, C. (1990) 'Barabaig Natural Resource Management: Sustainable Land Use under Threat of Destruction', United Nations Research Institute for Social Development (UNRISD) Discussion Paper No. 12.

Latorre, F. G., Williams, C., ten Kate, K. and Phyllida, C. (2001) *Results of the Pilot Project for Botanic Gardens: Principles on Access to Genetic Resources and Benefit Sharing.* Common Policy Guidelines to Assist with their Implementation and the Explanatory Text, Kew: Royal Botanic Gardens.

Lemley, M. A. (2005) 'The Changing Meaning of Patent Claim Terms', *Michigan Law Review,* 104 (1): 105–28.

Lerch, A. (1998) 'Property Rights and Biodiversity', *European Journal of Law and Economics,* 6 (3): 285–304.

Lewinski, S. von (ed.) (2004) *Indigenous Heritage and Intellectual Property: Genetic Resources, Traditional Knowledge and Folklore,* The Hague: Kluwer Law International.

Lewis, W. H. (2000) 'Ethnopharmacology and the Search for New Therapeutics', in P. E. Minnis and W. J. Elisens (eds), *Biodiversity and Native America,* Norman, Okla.: University of Oklahoma Press, pp. 74–96.

Lewis, W. H. and Ramani, V. (2003) 'Ethics and Practice in Ethnobiology: Analysis of the International Cooperative Biodiversity Project-Peru', paper presented at the International Conference on Biodiversity and Biotechnology and the Protection of Traditional Knowledge at Washington University, St Louis, 4–6 April.

Machlup, F. (1958) *An Economic Review of the Patent System,* Study No. 15 of the Sub-Committee on Patents, Trademarks, and Copyrights, of the Committee on the Judiciary, US Senate 85th Session, 2nd Session, Washington.
Machlup, F. and Penrose, E. (1950) 'The Patent Controversy in the Nineteenth Century', *Journal of Economic History,* 10 (1): 1–29.
Maffi, L. (ed.) (2001) *One Biocultural Diversity: Linking Languages, Knowledge and the Environment,* Washington, DC: Smithsonian Institution Press.
McManis, C. R. (1998) 'The Interface Between International Intellectual Property and Environmental Protection: Biodiversity and Biotechnology', *Washington University Law Quarterly,* 76 (1): 255–73.
Merges, R. P. (1995) 'The Economic Impact of Intellectual Property Rights: An Overview and Guide', *Journal of Cultural Economics,* 19 (2): 103–17.
Mgbeoji, I. (2006) *Global Biopiracy: Patents, Plants and Indigenous Knowledge,* Ithaca, NY: Cornell University Press and UBC Press.
Mooney, P. R. (2000) 'Why We Call it Biopiracy', in H. Svarstad and S. S. Dhilion (eds), *Responding to Bioprospecting: From Biodiversity in the South to Medicines in the North,* Oslo: Spartacus Forlag AS, pp. 38–55.
Morphy, H. (1991) *Ancestral Connections: Art and Aboriginal System of Knowledge,* Chicago, Ill.: University of Chicago Press.
Munoz Tellez, V. (2006) *Biodiversity in the Amazon: The 'Cupuaçu is Ours' Campaign in Brazil,* Intellectual Property and NGOs Project of Queen Mary, University of London. Available online at http://www.ipngos.org/NGO%20Briefings/The%20Case%20of% 20Cupuacu.pdf (accessed 9 January 2009).
Muralidharan, R. (1997) 'Everything on Patents', *WIPO Infotech Handbook on Industrial Property Law,* Geneva: World Intellectual Property Organization.
Newman, D. J., Cragg, G. M. and Snader, K. M. (2003) 'Natural Products as Sources of New Drugs over the Period 1981–2002', *Journal of Natural Products,* 66 (7): 1022–37.
Paramasivam, M. (2008) 'Quantative Determination of Curcuminoids in Turmeric Powder by HPLTC Technique', *Current Science*, 95 (11): 1529–31.
PhRMA (2003) 'Highlights from the Pharmaceutical Industry Profile: Prescription Medicine 25 Years Ago and Today, Changing Trends, Enduring Needs, New Medicine, New Hopes, Insights', available online at http://www.phrma.org/publications/publications/ 2003–10–07.892.pdf (accessed 30 January 2009).
Plotkin, M. J. (2000) *Medicine Quest: In Search of Nature's Healing Secrets,* New York: Viking.
Posey, D. A. (1995) 'The Working Group on Traditional Resources Rights', *Bulletin,* 1 (spring).
——(1996) *Traditional Resources Rights: International Instruments for the Protection and Compensation for Indigenous Peoples and Local Communities,* Gland: IUCN Biodiversity Programme.
Ragavan, S. (2001) 'Protection of Traditional Knowledge', *Minnesota Intellectual Property Review,* 2 (1): 1–60.
Reid, W. V., Laird, S. A., Meyer, C. A., Games, R., Sittenfeld, A., Janzen, D. H., Gollin, M. and Juma, C. (1993) *Biodiversity Prospecting: Using Genetic Resources for Sustainable Development,* Washington, DC: World Resources Institute.
Reisdorff, C., Gasparotto, L. and Lieberei, R. (2000) 'Development and Production of Cupuaçu Plants (*Theobroma grandiflorum*) on Degraded Areas: Implications of Water Relations', paper presented at the German-Brazilian Workshop on Neotropical Ecosystems: Achievements and Prospects of Cooperative Research Hamburg, 3–8 September.

Available online at http://www.biologie.uni-hamburg.de/bzf/oknu/proceedingsneotropecosys/p0661_reisdorff.pdf (accessed 9 January 2009).

Reyes-Garcia, V., Godoy, R., Vadez, V., Apaza, L., Byron, E., Huanca, T., Leonard, W. R., Perez, E. and Wilkie, D. (2003) 'Ethnobotanical Knowledge Shared Widely Among Tsimane Amerindians, Bolivia', *Science,* 299 (5613): 1707.

Sampath, P. G. (2005) *Regulating Bioprospecting Institutions for Drug Research Access and Benefit Sharing,* Tokyo: United Nations University Press.

Savigny, C. (1849) *System des Heutigen Romischen,* Vol. VIII, Berlin: Bei Veit und Comp.

Schmidlehner, M. (2003) 'Cupuaçu A Case of Amazonian Self-Assertion', available online at http://www.grain.org/seedling/?id=234 (accessed 9 January 2009).

Schuler, P. (2004) 'Biopiracy and the Commercialisation of Ethonotanical Knowledge', in J. M. Finger and P. Schuler (eds), *Poor Peoples' Knowledge: Promoting Intellectual Property in Developing Countries,* Oxford: World Bank and Oxford University Press, pp. 159–81.

Schultes, R. E. (1998) 'Primitive Plant Lore and Modern Conservation', *Orion Nature Quarterly* 7 (3): 8–15.

Sedjo, R. A. and Simpson, R. D. (1995) 'Property Rights, Externalities and Biodiversity', in T. Swanson (ed.), *The Economics and Ecology of Biodiversity Decline: The Forces Driving Global Change,* Cambridge: Cambridge University Press, pp. 79–88.

Sheldon, J. W., Balick, M. and Laird, S. A. (1997) 'Medicinal Plants: Can Utilization and Conservation Coexist?' *Advances in Economic Botany,* 12: 1–104.

Sherman, B. and Bently, L. (1999) *The Making of Modern Intellectual Property Law: The British Experience,* Cambridge: Cambridge University Press.

Simpson, T. (1997) *Indigenous Heritage and Self-Determination,* Copenhagen: International Work Group for Indigenous Affairs (IWGIA), No 86.

ten Kate, K. and Laird, S. A. (1999) *The Commercial Use of Biodiversity: Access to Genetic Resources and Benefit Sharing,* London: Earthscan.

Tewolde, B. G. E. (1996) 'The Convention on Biological Diversity, Intellectual Property Rights and the Interests of the South', in S. Tilahun and S. Edwards (eds), *The Movement for Collective Intellectual Rights,* Addis Ababa and London: Institute for Sustainable Development and Gaia Foundation.

Thornberry, P. (2002) *Indigenous Peoples and Human Rights,* Manchester: Manchester University Press.

Timmermans, K. (2003) 'Intellectual Property Rights and Traditional Medicine: Policy Dilemmas at the Interface', *Social Science and Medicine,* 57 (4): 745–56.

Tobin, B. and Swiderska, K. (2001) *Participation in Access and Benefit-Sharing Policy Case Study No. 2. Speaking in Tongues: Indigenous Participation in the Development of a Sui Generis Regime to Protect Traditional Knowledge in Peru,* London: International Institute for Environment and Development.

Tonye, M. M. (2004) 'Addressing the Concerns of Rural Communities About Access to Plants and Knowledge in a *Sui Generis* Legislation in Cameroon', *Journal of Bioscience,* 29 (4): 431–44.

——(2005) 'Biodiversity Regulatory Options: Involvement of Rural Communities in Decision-Making Processes in South Africa', *Journal of World Intellectual Property,* 8 (6): 809–24.

Twarog, S. and Kapoor, P. (eds) (2004) *Protecting and Promoting Traditional Knowledge: Systems, National Experiences and International Dimensions,* UNCTAD/DITC/TED/10, United Nations Conference for Trade and Development.

US National Research Council (1989) *Lost crop of the Incas: Little Known Plants of the Andes with Promise for Worldwide Cultivation*, Wasington, DC: National Academy Press.

Visser, C. J. (2004) 'Making Intellectual Property Laws Work for Traditional Knowledge', in J. M. Finger and P. Schuler (eds), *Poor People's Knowledge: Promoting Intellectual Property in Developing Countries,* Oxford: World Bank and Oxford University Press, pp. 207–40.

Weiner, A. B. (1992) *Inalienable Possessions: The Paradox of Keeping While Giving,* Berkeley, Calif.: University of California Press.

Wells, M. D. (1995) *Biodiversity Conservation and Local People's Development Aspirations: New Priorities for the 1990s,* Rural Development Forestry Network Series, No. 18d, London: ODI.

Wiser, G. M. and Downs, R. D. (1999) *Comments on Improving Identification of Prior Art: Recommendations on Traditional Knowledge Relating to Biological Diversity Submitted to the United States Patent and Trademark Office,* Washington, DC: Centre for International Environmental Law.

World Intellectual Property Organization (2001a) *Background Paper Prepared by the International Bureau.* WIPO Forum on Private International Law and Intellectual Property, Geneva, 30–1 January, WIPO/PIL/01/9.

——(2001b) *Intellectual Property Needs and Expectations of Traditional Knowledge Holders: WIPO Report on Fact-Finding Missions on Intellectual Property and Traditional Knowledge 1998–1999,* Geneva: World Intellectual Property Organization.

——(2001c) *Survey on Existing Forms of Intellectual Property Protection for Traditional Knowledge,* 8 August, WIPO/GRTKF/IC/2/5, Geneva: World Intellectual Property Organization.

——(2003) *Consolidated Survey of the Intellectual Property Protection of Traditional Knowledge,* WIPO/GRTKF/IC/5/7, Geneva: World Intellectual Property Organization.

——(2006a) *Options for Giving Effect to the International Dimension of the Committee's Work,* WIPO/GRTKF/IC/10, Geneva: World Intellectual Property Organization.

——(2006b) *The Protection of Traditional Cultural Expressions/Expressions of Folklore: Draft Objectives and Principles,* WIPO/GRTKF/IC/10/4, Geneva: World Intellectual Property Organization.

——(2006c) *The Protection of Traditional Knowledge: Draft Objectives and Principles,* WIPO/GRTKF/IC/10/5, Geneva: World Intellectual Property Organization.

World Rainforest Movement (2005) *Indigenous Peoples, their Forests, Struggles and Rights,* Montevideo: World Rainforest Movement.

World Wildlife Fund (2000) *Indigenous and Traditional Peoples of the World and Ecoregions Conservation: An Integrated Approach to Conserving the World's Biological and Cultural Diversity,* Gland: WWF International.

Younger, J. (1916) *British Thomson-Houston Co., Ltd.* v *Stonebridge Electrical Co., Ltd,* 33 RPC 166–1916.

Zerbe, N. (2005) 'Biodiversity, Ownership and Indigenous Knowledge: Exploring Legal Frameworks for Community, Farmers and Intellectual Property Rights in Africa', *Ecological Economics,* 53 (4): 493–513.

Zweifel, H. (1997) 'Biodiversity and the Appropriation of Women's Knowledge', *Indigenous Knowledge and Development Monitor,* 5 (1). Available online at http://www.nuffic.nl/ciran/ikdm/5–1/articles/zweifel.htm (accessed 11 October 2008).

Index

For Product Safety Concerns and Information please contact our EU representative GPSR@taylorandfrancis.com
Taylor & Francis Verlag GmbH, Kaufingerstraße 24, 80331 München, Germany

www.ingramcontent.com/pod-product-compliance
Ingram Content Group UK Ltd.
Pitfield, Milton Keynes, MK11 3LW, UK
UKHW020202200926
475702UK00004B/575
9780415479424